ELECTRICAL REFERENCES

2023 EDITION

Charles R. Miller

A Note from the Author...

Ugly's Electrical References is designed to be used as a quick on-the-job reference in the electrical industry. Used worldwide by electricians, engineers, contractors, designers, maintenance workers, instructors, and the military, Ugly's contains the most commonly required information in an easy-to-read format.

Ugly's Electrical References is not intended to be a substitute for the *National Electrical Code*®.

We salute the National Fire Protection Association for their dedication to the protection of lives and property from fire and electrical hazards through the sponsorship of the *National Electrical Code*®.

National Electrical Code®, *Standard for Electrical Safety in the Workplace*®, *NEC*®, and *70E*® are registered trademarks of the National Fire Protection Association, Quincy, MA 02169.

JONES & BARTLETT
LEARNING

TABLE OF CONTENTS

TABLE OF CONTENTS (continued)

TABLE OF CONTENTS (continued)

TABLE OF CONTENTS (continued)

TABLE OF CONTENTS (continued)

TABLE OF CONTENTS (continued)

<u>TABLE OF CONTENTS</u> (continued)

⚡ OHM'S LAW

Ohm's Law is the relationship between voltage (E), current (I), and resistance (R). The rate of the current flow is equal to electromotive force divided by resistance.

I = **Intensity of Current = Amperes**
E = **Electromotive Force = Volts**
R = **Resistance = Ohms**
P = **Power = Watts**

The three basic Ohm's Law formulas are:

$$I = \frac{E}{R} \qquad\qquad R = \frac{E}{I} \qquad\qquad E = I \times R$$

Below is a chart containing the formulas related to Ohm's Law. To use the chart, start in the center circle and select the value you need to find, I (amps), R (ohms), E (volts), or P (watts). Then select the formula containing the values you know from the corresponding chart quadrant.

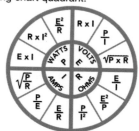

Example:

An electrical appliance is rated at 1200 watts and is connected to 120 volts. How much current will it draw?

$$\textbf{Amperes} = \frac{\textbf{Watts}}{\textbf{Volts}} \qquad I = \frac{P}{E} \qquad I = \frac{1200}{120} = 10 \text{ Amps}$$

What is the resistance of the same appliance?

$$\textbf{Ohms} = \frac{\textbf{Volts}}{\textbf{Amperes}} \qquad R = \frac{E}{I} \qquad R = \frac{120}{10} = 12 \text{ Ohms}$$

1

🔌 OHM'S LAW

In the preceding example, we know the following values:

I = amps = 10 Amps $\qquad$ R = ohms = 12 Ohms
E = volts = 120 Volts $\qquad$ P = watts = 1200 Watts

We can now see how the 12 formulas in the Ohm's Law chart can be applied.

Amps $= \sqrt{\dfrac{\text{Watts}}{\text{Ohms}}}$ $\qquad$ $I = \sqrt{\dfrac{P}{R}} = \sqrt{\dfrac{1200}{12}} = \sqrt{100} = 10$ Amps

Amps $= \dfrac{\text{Watts}}{\text{Volts}}$ $\qquad$ $I = \dfrac{P}{E} = \dfrac{1200}{120} = 10$ Amps

Amps $= \dfrac{\text{Volts}}{\text{Ohms}}$ $\qquad$ $I = \dfrac{E}{R} = \dfrac{120}{12} = 10$ Amps

Watts $= \dfrac{\text{Volts}^2}{\text{Ohms}}$ $\qquad$ $P = \dfrac{E^2}{R} = \dfrac{120^2}{12} = \dfrac{14400}{12} = 1200$ Watts

Watts $=$ **Volts** $\times$ **Amps** $\qquad$ $P = E \times I = 120 \times 10 = 1200$ Watts

Watts $=$ **Amps**^{2} $\times$ **Ohms** $\qquad$ $P = I^2 \times R = 10^2 \times 12 = 1200$ Watts

Volts $= \sqrt{\text{Watts} \times \text{Ohms}}$ $\qquad$ $E = \sqrt{P \times R} = \sqrt{1200 \times 12} = \sqrt{14400} = 120$ Volts

Volts $=$ **Amps** $\times$ **Ohms** $\qquad$ $E = I \times R = 10 \times 12 = 120$ Volts

Volts $= \dfrac{\text{Watts}}{\text{Amps}}$ $\qquad$ $E = \dfrac{P}{I} = \dfrac{1200}{10} = 120$ Volts

Ohms $= \dfrac{\text{Volts}^2}{\text{Watts}}$ $\qquad$ $R = \dfrac{E^2}{P} = \dfrac{120^2}{1200} = \dfrac{14400}{1200} = 12$ Ohms

Ohms $= \dfrac{\text{Watts}}{\text{Amps}^2}$ $\qquad$ $R = \dfrac{P}{I^2} = \dfrac{1200}{10^2} = 12$ Ohms

Ohms $= \dfrac{\text{Volts}}{\text{Amps}}$ $\qquad$ $R = \dfrac{E}{I} = \dfrac{120}{10} = 12$ Ohms

🔌 SERIES CIRCUITS

A series circuit is a circuit that has only one path through which the electrons may flow.

Rule 1: The total current in a series circuit is equal to the current in any other part of the circuit.

Total Current $I_T = I_1 = I_2 = I_3$, etc.

Rule 2: The total voltage in a series circuit is equal to the sum of the voltages across all parts of the circuit.

Total Voltage $E_T = E_1 + E_2 + E_3$, etc.

Rule 3: The total resistance of a series circuit is equal to the sum of the resistances of all the parts of the circuit.

Total Resistance $R_T = R_1 + R_2 + R_3$, etc.

Formulas from Ohm's Law

$$\text{Amperes} = \frac{\text{Volts}}{\text{Resistance}} \quad \text{or} \quad I = \frac{E}{R}$$

$$\text{Resistance} = \frac{\text{Volts}}{\text{Amperes}} \quad \text{or} \quad R = \frac{E}{I}$$

$$\text{Volts} = \text{Amperes} \times \text{Resistance} \quad \text{or} \quad E = I \times R$$

Example 1: Find the total voltage, total current, and total resistance of the following series circuit.

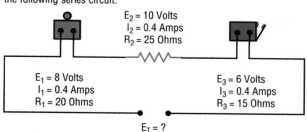

$E_2 = 10$ Volts
$I_2 = 0.4$ Amps
$R_2 = 25$ Ohms

$E_1 = 8$ Volts
$I_1 = 0.4$ Amps
$R_1 = 20$ Ohms

$E_3 = 6$ Volts
$I_3 = 0.4$ Amps
$R_3 = 15$ Ohms

$E_T = ?$
$I_T = ?$
$R_T = ?$

🔌 SERIES CIRCUITS

$$E_T = E_1 + E_2 + E_3$$
$$= 8 + 10 + 6$$
$$E_T = 24 \text{ Volts}$$

$$I_T = I_1 = I_2 = I_3$$
$$= 0.4 = 0.4 = 0.4$$
$$I_T = 0.4 \text{ Amps}$$

$$R_T = R_1 + R_2 + R_3$$
$$= 20 + 25 + 15$$
$$R_T = 60 \text{ Ohms}$$

Example 2: Find E_T, E_1, E_3, I_T, I_1, I_2, I_4, R_T, R_2, and R_4.
Remember that the total current in a series circuit is equal to the current in any other part of the circuit.

$E_1 = ?$ $E_3 = ?$
$I_1 = ?$ $I_3 = 0.5 \text{ Amps}$
$R_1 = 72 \text{ Ohms}$ $R_3 = 48 \text{ Ohms}$

$E_2 = 12 \text{ Volts}$ $E_4 = 48 \text{ Volts}$
$I_2 = ?$ $I_4 = ?$
$R_2 = ?$ $R_4 = ?$

$E_T = ?$ $I_T = ?$ $R_T = ?$

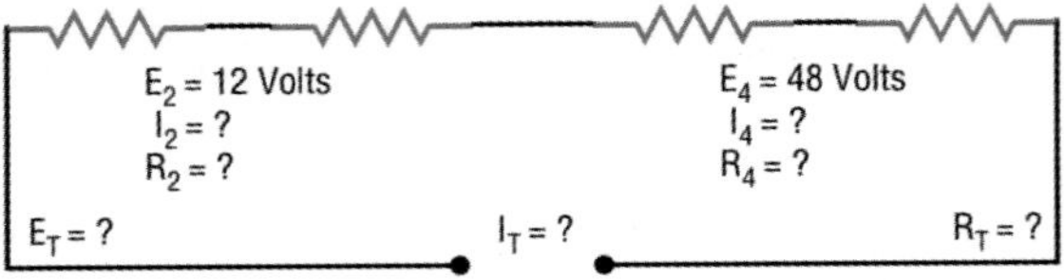

$$I_T = I_1 = I_2 = I_3 = I_4$$
$$I_T = I_1 = I_2 = 0.5 = I_4$$
$$0.5 = 0.5 = 0.5 = 0.5 = 0.5$$
$I_T = 0.5 \text{ Amps}$ $I_2 = 0.5 \text{ Amps}$
$I_1 = 0.5 \text{ Amps}$ $I_4 = 0.5 \text{ Amps}$

$$E_1 = I_1 \times R_1$$
$$= 0.5 \times 72$$
$$E_1 = 36 \text{ Volts}$$

$$E_T = E_1 + E_2 + E_3 + E_4$$
$$= 36 + 12 + 24 + 48$$
$$E_T = 120 \text{ Volts}$$

$$E_3 = I_3 \times R_3$$
$$= 0.5 \times 48$$
$$E_3 = 24 \text{ Volts}$$

$$R_T = R_1 + R_2 + R_3 + R_4$$
$$= 72 + 24 + 48 + 96$$
$$R_T = 240 \text{ Ohms}$$
$$R_2 = \frac{E_2}{I_2} = \frac{12}{0.5}$$
$$R_2 = 24 \text{ Ohms}$$

$$R_4 = \frac{E_4}{I_4} = \frac{48}{0.5}$$
$$R_4 = 96 \text{ Ohms}$$

4

⚡ PARALLEL CIRCUITS

A parallel circuit is a circuit that has more than one path through which the electrons may flow.

Rule 1: The total current in a parallel circuit is equal to the sum of the currents in all the branches of the circuit.
Total Current $I_T = I_1 + I_2 + I_3$, etc.

Rule 2: The total voltage across any branch in parallel is equal to the voltage across any other branch and is also equal to the total voltage.
Total Voltage $E_T = E_1 = E_2 = E_3$, etc.

Rule 3: The total resistance of a parallel circuit is found by applying Ohm's Law to the total values of the circuit.

$$\text{Total Resistance} = \frac{\text{Total Voltage}}{\text{Total Amperes}} \quad \text{or} \quad R_T = \frac{E_T}{I_T}$$

Example 1: Find the total current, total voltage, and total resistance of the following parallel circuit.

E_1 = 120 Volts	E_2 = 120 Volts	E_3 = 120 Volts
I_1 = 2 Amps	I_2 = 1.5 Amps	I_3 = 1 Amps
R_1 = 60 Ohms	R_2 = 80 Ohms	R_3 = 120 Ohms

$$
\begin{aligned}
I_T &= I_1 + I_2 + I_3 \\
&= 2 + 1.5 + 1 \\
I_T &= 4.5 \text{ Amps}
\end{aligned}
\qquad
\begin{aligned}
E_T &= E_1 = E_2 = E_3 \\
&= 120 = 120 = 120 \\
E_T &= 120 \text{ Volts}
\end{aligned}
$$

$$R_T = \frac{E_T}{I_T} = \frac{120 \text{ Volts}}{4.5 \text{ Amps}} = 26.66 \text{ Ohms Resistance}$$

Note: In a parallel circuit, the total resistance is always less than the resistance of any branch. If the branches of a parallel circuit have the same resistance, then each will draw the same current. If the branches of a parallel circuit have different resistances, then each will draw a different current. In either series or parallel circuits, the larger the resistance, the smaller the current drawn.

PARALLEL CIRCUITS

To determine the total resistance in a parallel circuit when the total current and total voltage are unknown:

$$\frac{1}{\text{Total Resistance}} = \frac{1}{R_1} + \frac{1}{R_2} + \frac{1}{R_3} + \dots \text{etc.}$$

Example 2: Find the total resistance.

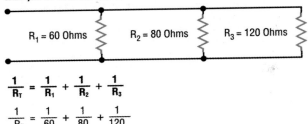

$$\frac{1}{R_T} = \frac{1}{R_1} + \frac{1}{R_2} + \frac{1}{R_3}$$

$$\frac{1}{R_T} = \frac{1}{60} + \frac{1}{80} + \frac{1}{120}$$

$$\frac{1}{R_T} = \frac{4 + 3 + 2}{240} = \frac{9}{240} \quad \text{Use lowest common denominator (240).}$$

For a review of Adding Fractions and Common Denominators, see Ugly's pages 152–154.

$$\frac{1}{R_T} \diagdown \diagup \frac{9}{240} \quad \text{Cross multiply.}$$

$$9 \times R_T = 1 \times 240 \quad \text{or} \quad 9R_T = 240$$

Divide both sides of the equation by 9.

$$R_T = 26.66 \text{ Ohms Resistance}$$

Note: The total resistance of a number of equal resistors in parallel is equal to the resistance of one resistor divided by the number of resistors.

$$\text{Total Resistance} = \frac{\text{Resistance of One Resistor}}{\text{Number of Resistors in Circuit}}$$

PARALLEL CIRCUITS

Formula:

$$R_T = \frac{R}{N}$$

Example 3: Find the total resistance.

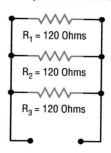

$R_1 = 120$ Ohms

$R_2 = 120$ Ohms

$R_3 = 120$ Ohms

There are three resistors in parallel. Each has a value of 120 Ohms resistance. According to the formula, if we divide the resistance of any one of the resistors by three, we will obtain the total resistance of the circuit.

$$R_T = \frac{R}{N} \quad \text{or} \quad R_T = \frac{120}{3}$$

Total Resistance = 40 Ohms

Note: To find the total resistance of only two resistors in parallel, multiply the resistances, and then divide the product by the sum of the resistors.

Formula: Total Resistance $= \dfrac{R_1 \times R_2}{R_1 + R_2}$

Example 4: Find the total resistance.

$R_1 = 40$ Ohms

$R_2 = 80$ Ohms

$$R_T = \frac{R_1 \times R_2}{R_1 + R_2}$$

$$= \frac{40 \times 80}{40 + 80}$$

$$R_T = \frac{3200}{120} = 26.66 \text{ Ohms}$$

7

COMBINATION CIRCUITS

In combination circuits, we combine series circuits with parallel circuits. Combination circuits make it possible to obtain the different voltages of series circuits and the different currents of parallel circuits.

Example 1: Parallel-Series Circuit.
Solve for all missing values.

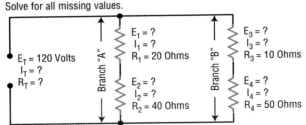

To solve:
1. Find the total resistance of each branch. Both branches are simple series circuits, so:

   $R_1 + R_2 = R_A$
   20 + 40 = 60 Ohms total resistance of branch "A"

   $R_3 + R_4 = R_B$
   10 + 50 = 60 Ohms total resistance of branch "B"

2. Redraw the circuit, combining resistors ($R_1 + R_2$) and ($R_3 + R_4$) so that each branch will have only one resistor.

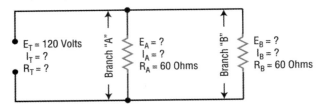

⚡ COMBINATION CIRCUITS

Note: We now have a simple parallel circuit, so:

$$E_T = E_A = E_B$$
$$120 \text{ Volts} = 120 \text{ Volts} = 120 \text{ Volts}$$

We now have a parallel circuit with only two resistors, and they are of equal value. We have a choice of three different formulas that can be used to find the total resistance of the circuit.

(1) $$R_T = \frac{R_A \times R_B}{R_A + R_B} = \frac{60 \times 60}{60 + 60} = \frac{3600}{120} = 30 \text{ Ohms}$$

(2) When the resistors of a parallel circuit are of equal value,

$$R_T = \frac{R}{N} = \frac{60}{2} = 30 \text{ Ohms} \qquad \text{or}$$

(3) $$\frac{1}{R_T} = \frac{1}{R_A} + \frac{1}{R_B} = \frac{1}{60} + \frac{1}{60} = \frac{2}{60} = \frac{1}{30}$$

$$\frac{1}{R_T} \diagup\!\!\!\diagdown \frac{1}{30} \quad \text{or} \quad 1 \times R_T = 1 \times 30 \quad \text{or} \quad R_T = 30 \text{ Ohms}$$

3. We know the values of E_T, R_T, E_A, R_A, E_B, R_B, R_1, R_2, R_3, and R_4. Next we will find the values of I_T, I_A, I_B, I_1, I_2, I_3, and I_4.

$$I_T = \frac{E_T}{R_T} \qquad \text{or} \quad \frac{120}{30} = 4 \qquad I_T = 4 \text{ Amps}$$

$$I_A = \frac{E_A}{R_A} \qquad \text{or} \quad \frac{120}{60} = 2 \qquad I_A = 2 \text{ Amps}$$

$$I_A = I_1 = I_2 \qquad \text{or} \quad 2 = 2 = 2 \qquad \begin{aligned} I_1 &= 2 \text{ Amps} \\ I_2 &= 2 \text{ Amps} \end{aligned}$$

$$I_B = \frac{E_B}{R_B} = \qquad \text{or} \quad \frac{120}{60} = 2 \qquad I_B = 2 \text{ Amps}$$

$$I_B = I_3 = I_4 \qquad \text{or} \quad 2 = 2 = 2 \qquad \begin{aligned} I_3 &= 2 \text{ Amps} \\ I_4 &= 2 \text{ Amps} \end{aligned}$$

⚡ COMBINATION CIRCUITS

4. We know that resistors #1 and #2 of branch "A" are in series. We
    also know that resistors #3 and #4 of branch "B" are in series.
    We have determined that the total current of branch "A" is 2 amps,
    and the total current of branch "B" is 2 amps. By using the series
    formula, we can find the current of each branch.

    Branch "A"

    $I_A = I_1 = I_2$

    $2 = 2 = 2$

    $I_1 = 2$ Amps

    $I_2 = 2$ Amps

    Branch "B"

    $I_B = I_3 = I_4$

    $2 = 2 = 2$

    $I_3 = 2$ Amps

    $I_4 = 2$ Amps

5. We were given the resistance values of all resistors.
    $R_1 = 20$ Ohms, $R_2 = 40$ Ohms, $R_3 = 10$ Ohms, and $R_4 = 50$ Ohms.
    By using Ohm's Law, we can determine the voltage drop across
    each resistor.

    $E_1 = R_1 \times I_1$

    $= 20 \times 2$

    $E_1 = 40$ Volts

    $E_2 = R_2 \times I_2$

    $= 40 \times 2$

    $E_2 = 80$ Volts

    $E_3 = R_3 \times I_3$

    $= 10 \times 2$

    $E_3 = 20$ Volts

    $E_4 = R_4 \times I_4$

    $= 50 \times 2$

    $E_4 = 100$ Volts

Example 2: Series Parallel Circuit.
Solve for all missing values.

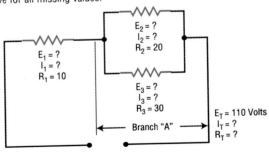

$E_2 = ?$
$I_2 = ?$
$R_2 = 20$

$E_1 = ?$
$I_1 = ?$
$R_1 = 10$

$E_3 = ?$
$I_3 = ?$
$R_3 = 30$

Branch "A"

$E_T = 110$ Volts
$I_T = ?$
$R_T = ?$

10

⚡ COMBINATION CIRCUITS

To solve:

1. We can see that resistors #2 and #3 are in parallel, and combined they are branch "A." When there are only two resistors, we use the following formula:

$$R_A = \frac{R_2 \times R_3}{R_2 + R_3} \quad \text{or} \quad \frac{20 \times 30}{20 + 30} \quad \text{or} \quad \frac{600}{50} \quad \text{or} \quad 12 \text{ Ohms}$$

2. We can now redraw our circuit as a simple series circuit.

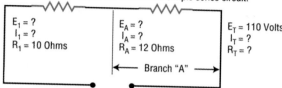

$E_1 = ?$
$I_1 = ?$
$R_1 = 10$ Ohms

$E_A = ?$
$I_A = ?$
$R_A = 12$ Ohms

◄──── Branch "A" ────►

$E_T = 110$ Volts
$I_T = ?$
$R_T = ?$

3. In a series circuit,
   $R_T = R_1 + R_A$ or $R_T = 10 + 12$ or 22 Ohms
   By using Ohm's Law,

   $$I_T = \frac{E_T}{R_T} = \frac{110}{22} = 5 \text{ Amps}$$

   In a series circuit,
   $I_T = I_1 = I_A$ or $I_T = 5$ Amps, $I_1 = 5$ Amps, and $I_A = 5$ Amps

   By using Ohm's Law,
   $E_1 = I_1 \times R_1 = 5 \times 10 = 50$ Volts
   $E_T - E_1 = E_A$ or $110 - 50 = 60$ Volts $= E_A$

   In a parallel circuit,
   $E_A = E_2 = E_3$ or $E_A = 60$ Volts
   $E_2 = 60$ Volts, and $E_3 = 60$ Volts

   By using Ohm's Law,

   $$I_2 = \frac{E_2}{R_2} = \frac{60}{20} = 3 \text{ Amps}$$

   $$I_3 = \frac{E_3}{R_3} = \frac{60}{30} = 2 \text{ Amps}$$

11

⚡ COMBINATION CIRCUITS

Problem:

Find the total resistance.
Redraw circuit as many times as necessary.
Correct answer is 100 Ohms.

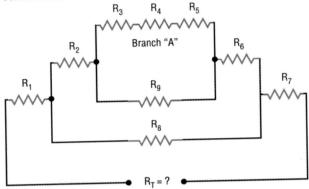

Given Values:

R_1	=	15 Ohms		
R_2	=	35 Ohms		
R_3	=	50 Ohms		
R_4	=	40 Ohms		
R_5	=	30 Ohms		

R_6	=	25 Ohms
R_7	=	10 Ohms
R_8	=	300 Ohms
R_9	=	60 Ohms

COMMON ELECTRICAL DISTRIBUTION SYSTEMS

120/240-Volt, Single-Phase, Three-Wire System

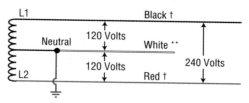

† · **Line one** ungrounded conductor colored **black**
† · **Line two** ungrounded conductor colored **red**
** · Grounded neutral conductor colored **white** or gray

120/240-Volt, Three-Phase, Four-Wire System (Delta High Leg)

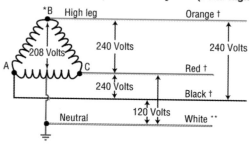

† · **A** phase ungrounded conductor colored **black**
†*· **B** phase ungrounded conductor colored **orange** or tagged
(high leg). (Caution: 208 volts orange to white)
† · **C** phase ungrounded conductor colored **red**
** · Grounded conductor colored **white** or gray (center tap)

** Grounded conductors are required to be white or gray or three white or gray stripes on
other than green insulation. See *NEC* 200.6.
* B phase of delta high leg must be orange or tagged. See *NEC* 110.15.
† Ungrounded conductor colors may be other than shown; see local ordinances or
specifications.

COMMON ELECTRICAL DISTRIBUTION SYSTEMS

120/208-Volt, Three-Phase, Four-Wire System (Wye Connected)

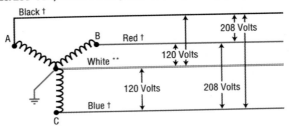

† · **A** phase ungrounded conductor colored **black**
† · **B** phase ungrounded conductor colored **red**
† · **C** phase ungrounded conductor colored **blue**
** · Grounded neutral conductor colored **white** or gray

277/480-Volt, Three-Phase, Four-Wire System (Wye Connected)

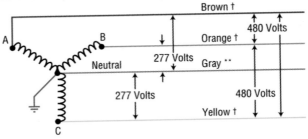

† · **A** phase ungrounded conductor colored **brown**
† · **B** phase ungrounded conductor colored **orange**
† · **C** phase ungrounded conductor colored **yellow**
** · Grounded neutral conductor colored **gray**

** Grounded conductors are required to be white or gray or three white or gray stripes on other than green insulation. See *NEC* 200.6.

† Ungrounded conductor colors may be other than shown; see local ordinances or specifications.

ELECTRICAL FORMULAS FOR CALCULATING AMPERES, HORSEPOWER, KILOWATTS, AND KVA

To Find	Direct Current	Alternating Current		
		Single Phase	Two Phase/Four Wire	Three Phase
Amperes when "HP" is known	$\dfrac{HP \times 746}{E \times \%EFF}$	$\dfrac{HP \times 746}{E \times \%EFF \times PF}$	$\dfrac{HP \times 746}{E \times \%EFF \times PF \times 2}$	$\dfrac{HP \times 746}{E \times \%EFF \times PF \times 1.73}$
Amperes when "kW" is known	$\dfrac{kW \times 1000}{E}$	$\dfrac{kW \times 1000}{E \times PF}$	$\dfrac{kW \times 1000}{E \times PF \times 2}$	$\dfrac{kW \times 1000}{E \times PF \times 1.73}$
Amperes when "kVA" is known		$\dfrac{kVA \times 1000}{E}$	$\dfrac{kVA \times 1000}{E \times 2}$	$\dfrac{kVA \times 1000}{E \times 1.73}$
Kilowatts (True power)	$\dfrac{E \times I}{1000}$	$\dfrac{E \times I \times PF}{1000}$	$\dfrac{E \times I \times PF \times 2}{1000}$	$\dfrac{E \times I \times PF \times 1.73}{1000}$
Kilovolt-Amperes "kVA" (Apparent power)		$\dfrac{E \times I}{1000}$	$\dfrac{E \times I \times 2}{1000}$	$\dfrac{E \times I \times 1.73}{1000}$
Horsepower	$\dfrac{E \times I \times \%EFF}{746}$	$\dfrac{E \times I \times \%EFF \times PF}{746}$	$\dfrac{E \times I \times \%EFF \times PF \times 2}{746}$	$\dfrac{E \times I \times \%EFF \times PF \times 1.73}{746}$

Percent Efficiency = % EFF = $\dfrac{Output (Watts)}{Input (Watts)}$ Power Factor = PF = $\dfrac{Power\ Used\ (Watts)}{Apparent\ Power}$ = $\dfrac{kW}{kVA}$

Note: Direct-current formulas do not use (PF, 2 or 1.73).
Single-phase formulas do not use (2 or 1.73).
Two-phase/four-wire formulas do not use (1.73).
Three-phase formulas do not use (2).

E = Volts
I = Amperes
W = Watts

15

⚡ TO FIND AMPERES

Direct Current

A.   When *horsepower* is known:

$$\text{Amperes} = \frac{\text{Horsepower x 746}}{\text{Volts x Efficiency}} \quad \text{or} \quad I = \frac{HP \times 746}{E \times \%EFF}$$

What current will a travel-trailer toilet draw when equipped with a 12-volt direct-current, ⅛-HP motor that has a 96% efficiency rating?

$$I = \frac{HP \times 746}{E \times \%EFF} = \frac{746 \times ⅛}{12 \times 0.96} = \frac{93.25}{11.52} = 8.09 \text{ Amps}$$

B.   When *kilowatts* are known:

$$\text{Amperes} = \frac{\text{Kilowatts x 1000}}{\text{Volts}} \quad \text{or} \quad I = \frac{KW \times 1000}{E}$$

A 75-kW, 240-volt direct-current generator is used to power a variable-speed conveyor belt at a rock-crushing plant. Determine the current.

$$I = \frac{KW \times 1000}{E} = \frac{75 \times 1000}{240} = 312.5 \text{ Amps}$$

Single Phase

A.   When *watts*, *volts*, and *power factor* are known:

$$\text{Amperes} = \frac{\text{Watts}}{\text{Volts x Power Factor}} \quad \text{or} \quad \frac{P}{E \times PF}$$

Determine the current when a circuit has a 1500-watt load, a power factor of 86%, and operates from a single-phase, 230-volt source.

$$I = \frac{1500}{230 \times 0.86} = \frac{1500}{197.8} = 7.58 \text{ Amps}$$

 **TO FIND AMPERES**

Single Phase (continued)

B. When *horsepower* is known:

$$\text{Amperes} = \frac{\text{Horsepower} \times 746}{\text{Volts} \times \text{Efficiency} \times \text{Power Factor}}$$

Determine the amp-load of a single-phase, ½-HP, 115-volt motor. The motor has an efficiency rating of 92% and a power factor of 80%.

$$I = \frac{\text{HP} \times 746}{\text{E} \times \%\text{EFF} \times \text{PF}} = \frac{½ \times 746}{115 \times 0.92 \times 0.80} = \frac{373}{84.64}$$

$I = 4.4$ Amps

C. When *kilowatts* are known:

$$\text{Amperes} = \frac{\text{Kilowatts} \times 1000}{\text{Volts} \times \text{Power Factor}} \quad \text{or} \quad I = \frac{\text{KW} \times 1000}{\text{E} \times \text{PF}}$$

A 230-volt, single-phase circuit has a 12-kW power load, and operates at 84% power factor. Determine the current.

$$I = \frac{\text{KW} \times 1000}{\text{E} \times \text{PF}} = \frac{12 \times 1000}{230 \times 0.84} = \frac{12000}{193.2} = 62 \text{ Amps}$$

D. When *kilovolt-amperes* are known:

$$\text{Amperes} = \frac{\text{Kilovolt-Amperes} \times 1000}{\text{Volts}} \quad \text{or} \quad I = \frac{\text{KVA} \times 1000}{\text{E}}$$

A 115-volt, 2-kVA, single-phase generator operating at full load will deliver 17.4 amps. (Prove.)

$$I = \frac{2 \times 1000}{115} = \frac{2000}{115} = 17.4 \text{ Amps}$$

Remember: By definition, amperes is the rate of current flow.

 **TO FIND AMPERES**

Three Phase

A.   When *Watts, Volts*, and *power Factor* are known:

$$\text{Amperes} = \frac{\text{Watts}}{\text{Volts x Power Factor x 1.73}}$$

$$\text{or} \qquad I = \frac{P}{E \text{ x PF x 1.73}}$$

Determine the current when a circuit has a 1500-watt load, a power factor of 86%, and operates from a three-phase, 230-volt source.

$$I = \frac{P}{E \text{ x PF x 1.73}} = \frac{1500}{230 \text{ x } 0.86 \text{ x } 1.73} = \frac{1500}{342.2}$$

$I = 4.4$ Amps

B.   When *horsepower* is Known:

$$\text{Amperes} = \frac{\text{Horsepower x 746}}{\text{Volts x Efficiency x Power Factor x 1.73}}$$

$$\text{or} \qquad I = \frac{\text{HP x 746}}{E \text{ x \%EFF x PF x 1.73}}$$

Determine the current draw of a three-phase, ½-HP 230-volt motor. The motor has an efficiency rating of 92% and a power factor of 80%.

$$I = \frac{\text{HP x 746}}{E \text{ x \%EFF x PF x 1.73}} = \frac{\text{½ x 746}}{230 \text{ x } 0.92 \text{ x } 0.80 \text{ x } 1.73}$$

$$= \frac{373}{293} = 1.27 \text{ Amps}$$

⚡ TO FIND AMPERES

Three Phase (continued)

C.   When *kilowatts* are known:

$$\text{Amperes} = \frac{\text{Kilowatts x 1000}}{\text{Volts x Power Factor x 1.73}}$$

$$\text{or} \qquad I = \frac{\text{KW X 1000}}{\text{E x PF x 1.73}}$$

A 230-volt, three-phase circuit, has a 12-KW power load and operates at 84% power factor. Determine the current.

$$I = \frac{\text{KW x 1000}}{\text{E x PF x 1.73}} = \frac{12 \times 1000}{230 \times 0.84 \times 1.73} = \frac{12000}{334.24}$$

$$I = 35.9 \text{ Amps}$$

D.   When *kilovolt-amperes* are known:

$$\text{Amperes} = \frac{\text{Kilovolt-Ampere x 1000}}{\text{E x 1.73}} = \frac{\text{KVA x 1000}}{\text{E x 1.73}}$$

A 230-volt, 4-KVA, three-phase generator operating at full load will deliver 10 amps. (Prove.)

$$I = \frac{\text{KVA x 1000}}{\text{E x 1.73}} = \frac{4 \times 1000}{230 \times 1.73} = \frac{4000}{397.9}$$

$$I = 10 \text{ Amps}$$

🔩 TO FIND HORSEPOWER

Direct Current

$$\text{Horsepower} = \frac{\text{Volts x Amperes x Efficiency}}{746}$$

A 12-volt motor draws a current of 8.1 amps and has an efficiency rating of 96%. Determine the horsepower.

$$\text{HP} = \frac{\text{E x I x \%EFF}}{746} = \frac{12 \times 8.09 \times 0.96}{746} = \frac{93.19}{746}$$

$$\text{HP} = 0.125 = \tfrac{1}{8} \text{ HP} \ (\tfrac{1}{8} = 1 \div 8 = 0.125)$$

Single Phase

$$\text{HP} = \frac{\text{Volts x Amperes x Efficiency x Power Factor}}{746}$$

A single-phase, 115-volt ac motor has an efficiency rating of 92% and a power factor of 80%. Determine the horsepower if the amp-load is 4.4 amps.

$$\text{HP} = \frac{\text{E x I x \%EFF x PF}}{746} = \frac{115 \times 4.4 \times 0.92 \times 0.80}{746}$$

$$\text{HP} = \frac{372.416}{746} = 0.4992 = \tfrac{1}{2} \text{ HP}$$

Three Phase

$$\text{HP} = \frac{\text{Volts x Amperes x Efficiency x Power Factor x 1.73}}{746}$$

A three-phase, 480-volt motor draws a current of 52 amps. The motor has an efficiency rating of 94% and a power factor of 80%. Determine the horsepower.

$$\text{HP} = \frac{\text{E x I x \%EFF x PF x 1.73}}{746} = \frac{460 \times 52 \times 0.94 \times 0.80 \times 1.73}{746}$$

$$\text{HP} = 41.7 \text{ HP}$$

🔌 TO FIND WATTS

The electrical power m any part of a ctrcurt is equal to the current in that part multiplied by the voltage across that part of the circuit.

A watt is the power used when 1 volt causes 1 amps to flow in a circuit.

One horsepower is the amount of energy required to lift 33000 pounds, 1 foot, in 1 minute. The electrical equivalent of 1 HP is 745.6 watts. One watt is the amount of energy required to lift 44.26 pounds, 1 foot, in 1 minute, Watts is power, and power is the amount of work done in a given time.

When *volts* and *amperes* are known:

Power (Watts) = Volts x Amperes

A 120-volt ac circuit draws a current of 5 amps. Determine the power consumption.

$$P = E \times I = 120 \times 5 = 600 \text{ Watts}$$

Now determine the resistance of this circuit.

$$\text{Resistanee} = \frac{\text{Watts}}{\text{Amps}^2}$$

$$R = \frac{P}{I^2} = \frac{600}{5 \times 5} = \frac{600}{25} = 24 \text{ Ohms}$$

or

$$\text{Resistanee} = \frac{\text{Volts}^2}{\text{Watts}} \quad \text{or} \quad R = \frac{E^2}{P}$$

$$R = \frac{120 \times 120}{600} = \frac{14400}{600} = 24 \text{ Ohms}$$

Note: Refer to the formulas of the Ohm's Law chart on page 1.

 # TO FIND KILOWATTS

Direct Current

$$\text{Kilowatts} = \frac{\text{Volts} \times \text{Amperes}}{1000}$$

A 120-volt dc motor draws a current of 40 amps.
Determine the kilowatts.

$$\text{KW} = \frac{\text{E} \times \text{I}}{1000} = \frac{120 \times 40}{1000} = \frac{4800}{1000} = 4.8 \text{ KW}$$

Single Phase

$$\text{Kilowatts} = \frac{\text{Volts} \times \text{Amperes} \times \text{Power Factor}}{1000}$$

A single-phase, 115-volt ac motor draws a current of 20 amps and has a power factor rating of 86%. Determine the kilowatts.

$$\text{KW} = \frac{\text{E} \times \text{I} \times \text{PF}}{1000} = \frac{115 \times 20 \times 0.86}{1000} = \frac{1978}{1000} = 1.978 = 2 \text{ KW}$$

Three Phase

$$\text{Kilowatts} = \frac{\text{Volts} \times \text{Amperes} \times \text{Power Factor} \times 1.73}{1000}$$

A three-phase, 466-volt ac motor draws a current of 52 amps and has a power factor rating of 80%. Determine the kilowatts.

$$\text{KW} = \frac{\text{E} \times \text{I} \times \text{PF} \times 1.73}{1000} = \frac{460 \times 52 \times 0.80 \times 1.73}{1000}$$

$$= \frac{33105}{1000} = 33.105 = 33 \text{ KW}$$

🔌 TO FIND KILOVOLT-AMPERES

Single Phase

$$\text{Kilovolt-Amperes} = \frac{\text{Volts x Amperes}}{1000}$$

A single-phase, 240-volt generator delivers 41.66 amps at full load. Determine the kilovolt-amperes rating

$$\text{KVA} = \frac{\text{E x I}}{1000} = \frac{240 \times 41.66}{1000} = \frac{9998.4}{1000} = 10 \text{ KVA}$$

Three Phase

$$\text{Kilovolt-Amperes} = \frac{\text{Volts x Amperes x 1.73}}{1000}$$

A three-phase, 460-volt generator delivers 52 amps. Determine the kilovolt-amperes rating.

$$\text{KVA} = \frac{\text{E x I x 1.73}}{1000} = \frac{460 \times 52 \times 1.73}{1000} = \frac{41382}{1000}$$

$$= 41.382 = 41 \text{ KVA}$$

Note: KVA = Apparent Power = Power Before Used, such as the rating of a transformer.

Kirchhoff's Laws

First Law (Current):
The sum of the currents arriving at any point in a circuit must equal the sum of the currents leaving that point.

Second Law (Voltage):
The total voltage applied to any closed circuit path is always equal to the sum of the voltage drops in that path.

or

The algebraic sum of all the voltages encountered in any loop equals zero.

🔌 TO FIND CAPACITANCE

Capacitance (C)

$$C = \frac{Q}{E} \quad \text{or} \quad \text{Capacitance} = \frac{\text{Coulombs}}{\text{Volts}}$$

Capacitance is the property of a circuit or body that permits it to store an electrical charge equal to the accumulated charge divided by the voltage. Capacitance is expressed in farads.

A. To determine the total capacity of capacitors and/or condensers connected in series:

$$\frac{1}{C_T} = \frac{1}{C_1} + \frac{1}{C_2} + \frac{1}{C_3} + \frac{1}{C_4}$$

Determine the total capacity of four 12-microfarad capacitors connected in series

$$\frac{1}{C_T} = \frac{1}{C_1} + \frac{1}{C_2} + \frac{1}{C_3} + \frac{1}{C_4}$$

$$= \frac{1}{12} + \frac{1}{12} + \frac{1}{12} + \frac{1}{12} = \frac{4}{12}$$

$$\frac{1}{C_T} \diagdown \diagup \frac{4}{12} \quad \text{or} \quad C_1 \times 4 = 12 \quad \text{or} \quad C_T = \frac{12}{4} = 3 \text{ Microfarads}$$

B. To determine the total capacity of capacitors and/or condensers connected in parallel:

$$C_T = C_1 + C_2 + C_3 + C_4$$

Determine the total capacity of four 12-microfarad capacitors connected in parallel:

$$C_T = C_1 + C_2 + C_3 + C_4$$

$$C_T = 12 + 12 + 12 + 12 = 48 \text{ Microfarads}$$

A farad is the unit of capacitance of a condenser that retains 1 coulomb of charge with 1 volt difference of potential.

1 farad = 1000000 microfarads

SIX-DOT COLOR CODE FOR MICA AND MOLDED PAPER CAPACITORS

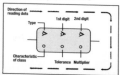

Type	Color	1st Digit	2nd Digit	Multiplier	Tolerance (%)	Characteristic or class
JAN, Mica	Black	0	0	1	± 1	
	Brown	1	1	10	± 2	
	Red	2	2	100	± 3	
	Orange	3	3	1000	± 4	
	Yellow	4	4	10000	± 5	Applies to
	Green	5	5	100000	± 6	Temperature
	Blue	6	6	1000000	± 7	Coefficient
	Violet	7	7	10000000	± 8	or Methods
	Gray	8	8	100000000	± 9	of Testing
ETA, Mica	White	9	9	1000000000		
	Gold			0.1	± 10	
Molded Paper	Silver			0.01	± 20	
	Body					

RESISTOR COLOR CODE

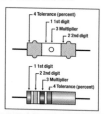

Color	1st Digit	2nd Digit	Multiplier	Tolerance (%)
Black	0	0	1	
Brown	1	1	10	
Red	2	2	100	
Orange	3	3	1000	
Yellow	4	4	10000	
Green	5	5	100000	
Blue	6	6	1000000	
Violet	7	7	10000000	
Gray	8	8	100000000	
White	9	9	1000000000	
Gold			0.1	± 5%
Silver			0.01	± 10%
No Color				± 20%

MAXIMUM PERMISSIBLE CAPACITOR KVAR FOR USE WITH OPEN-TYPE, THREE-PHASE, 60-CYCLE INDUCTION MOTORS

Motor Rating HP	3600 RPM		1800 RPM		1200 RPM	
	Maximum Capacitor Rating KVAR	Reduction in Line Current %	Maximum Capacitor Rating KVAR	Reduction in Line Current %	Maximum Capacitor Rating KVAR	Reduction in Line Current %
10	3	10	3	11	3.5	14
15	4	9	4	10	5	13
20	5	9	5	10	6.5	12
25	6	9	6	10	7.5	11
30	7	8	7	9	9	11
40	9	8	9	9	11	10
50	12	8	11	9	13	10
60	14	8	14	8	15	10
75	17	8	16	8	18	10
100	22	8	21	8	25	9
125	27	8	26	8	30	9
150	32.5	8	30	8	35	9
200	40	8	37.5	8	42.5	9

Motor Rating HP	900 RPM		720 RPM		600RPM	
	Maximum Capacitor Rating KVAR	Reduction in Line Current %	Maximum Capacitor Rating KVAR	Reduction in Line Current %	Maximum Capacitor Rating KVAR	Reduction in Line Current %
10	5	21	6.5	27	7.5	31
15	6.5	18	8	23	9.5	27
20	7.5	16	9	21	12	25
25	9	15	11	20	14	23
30	10	14	12	18	16	22
40	12	13	15	16	20	20
50	15	12	19	15	24	19
60	18	11	22	15	27	19
75	21	10	26	14	32.5	18
100	27	10	32.5	13	40	17
125	32.5	10	40	13	47.5	16
150	37.5	10	47.5	12	52.5	15
200	47.5	10	60	12	65	14

Note: If capacitors of a lower rating than the values given in the table are used, the percentage reduction in line current given in the table should be reduced proportionately

POWER FACTOR CORRECTION

Table Values × **Kilowat of Capacitors Needed to Correct from Existing to Desired Power Factor**

Existing power factor %	Corrected power factor					
	100%	95%	90%	85%	80%	75%
50	1.732	1.403	1.247	1.112	0.982	0.850
52	1.643	1.314	1.158	1.023	0.893	0.761
54	1.558	1.229	1.073	0.938	0.808	0.676
55	1.518	1.189	1.033	0.898	0.768	0.636
56	1.479	1.150	0.994	0.859	0.729	0.597
58	1.404	1.075	0.919	0.784	0.654	0.522
60	1.333	1.004	0.848	0.713	0.583	0.451
62	1.265	0.936	0.780	0.645	0.515	0.383
64	1.201	0.872	0.716	0.581	0.451	0.319
65	1.168	0.839	0.683	0.548	0.418	0.286
66	1.139	0.810	0.654	0.519	0.389	0.257
68	1.078	0.749	0.593	0.458	0.328	0.196
70	1.020	0.691	0.535	0.400	0.270	0.138
72	0.964	0.635	0.479	0.344	0.214	0.082
74	0.909	0.580	0.424	0.289	0.159	0.027
75	0.882	0.553	0.397	0.262	0.132	
76	0.855	0.526	0.370	0.235	0.105	
78	0.802	0.473	0.317	0.182	0.052	
80	0.750	0.421	0.265	0.130		
82	0.698	0.369	0.213	0.078		
84	0.646	0.317	0.161			
85	0.620	0.291	0.135			
86	0.594	0.265	0.109			
88	0.540	0.211	0.055			
90	0.485	0.156				
92	0.426	0.097				
94	0.363	0.034				
95	0.329					

Typical Problem. With a load of 500 KW at 70% power factor, it is desired to find the KVA of capacitors required to correct the power factor to 85%.
Solution: From the table, select the multiplying factor 0.400 corresponding to the existing 70%, and the corrected 85% power factor.
0.400 x 500 = 200 KVA of capacitors required

⚡ POWER FACTOR AND EFFICIENCY EXAMPLE

A squirrel cage induction motor is rated 10 HP, 208 volt, three-phase and has a nameplate rating of 27.79 amps. A wattmeter reading indicates 8 kilowatts of consumed (true) power. Calculate apparent power (KVA), power factor, efficiency, internal losses, and size of the capacitor in kilovolts reactive reactive (KVAR) needed to correct tie power factor to unity (100%).

Apparent input power: kilovolt-amperes (kVA)
KVA = (E x I x 1.73) / 1000 = (208 x 27.79 x 1.73) /1000 = **10 kVA**

Power factor (PF) = ratio of true power (kW) to apparent power (kVA)
Kilowatts/kilovolt-amperes =8 KW/10 kVA =0.8= **80% Power Factor**
80% of the 10-KVA apparent power input performs work.

Motor output in kilowatts =10 HP x 746 watts=7460 watts=**7.46 kW**
Efficiency=watts out/watts in = 7.46 kW/8 kW=0.9325 =**93.25% efficiency**

Internal losses (heat, friction, hysteresis) = 8 KW–7.46 KW = **0.54 kW** (540 watts)

Kilovolt-amperes reactive (KVAR)   (Power stored in motor magnetic field)
$$KVAR = \sqrt{KVA^2 - KW^2} = \sqrt{10\ KVA^2 - 8\ KW^2} = \sqrt{100 - 64} = \sqrt{36} = \textbf{6 KVAR}$$

The size capacitor needed to equal the motor's stored reactive power is 6 KVAR. (A capacitor stores reactive power in its electrostatic field.)

The power source must supply the current to perform work and maintain the motor's magnetic field. Before power factor correction, tins was 27.79 amperes. The motor magnetizing current after power factor correction is supplied by circulation of current between the motor and the electrostatic field of the capacitor and is no longer supplied by power source after initial startup. The motor feeder current after correction to 100% will equal the amount required by the input watts in this case (8 kW)/(208 volts x 1.73)= (8 x 1000)/(208 volts x 1.73) =**22.2 amps.**

- Kilo = 1000. For example: 1000 Watts = 1 Kilowatt.
- Inductive loads (motors, cods) have lagging currents, and capacitive loads have leading currents.
- Inductance and capacitance have opposite effects in a circuit and can cancel each other.

28

🔌 TO FIND INDUCTANCE

Inductance (L)

Inductance is the production of magnetization of electrification in a body by the proximity of a magnetic field or electric charge, or of the electric current in a conductor by the variation of the magnetic field in its vicinity. The unit of measurement for inductance is the henry (H).

A. To find the total inductance of coils connected in series:

$$L_T = L_1 + L_2 + L_3 + L_4$$

Determine the total inductance of four coils connected in series. Each coil has an inductance of 4 Henries

$$L_T = L_1 + L_2 + L_3 + L_4$$

$$= 4 + 4 + 4 + 4 = 16 \text{ Henries}$$

B. To find the total inductance of coils connected in parallel:

$$\frac{1}{L_T} = \frac{1}{L_1} + \frac{1}{L_2} + \frac{1}{L_3} + \frac{1}{L_4}$$

Determine the total inductance of four coils connected in parallel. Each coil has an inductance of 4 Henries

$$\frac{1}{L_T} = \frac{1}{L_1} + \frac{1}{L_2} + \frac{1}{L_3} + \frac{1}{L_4}$$

$$\frac{1}{L_T} = \frac{1}{4} + \frac{1}{4} + \frac{1}{4} + \frac{1}{4}$$

$$\frac{1}{L_T} = \frac{4}{4} \quad \text{or } L_T \times 4 = 1 \times 4 \quad \text{or} \quad L_T = \frac{4}{4} = 1 \text{ Henry}$$

An induction coil is a device consisting of two concentric coils and an interrupter, which changes a low steady voltage into a high intermittent alternating voltage by electromagnetic induction. Most often used as a spark coil.

🔲 TO FIND IMPEDANCE

Impedance (Z)

Impedance is the total opposition to an alternating current presented by a circuit. Expressed in ohms.

A. When *volts* and *amperes* are known:

$$\text{Impedance} = \frac{\text{Volts}}{\text{Amperes}} \quad \text{or} \quad Z = \frac{E}{I}$$

Determine the impedance of a 120-volt ac circuit that draws a current of 4 amps.

$$Z = \frac{E}{I} = \frac{120}{4} = 30 \text{ Ohms}$$

B. When *resistance* and *reactance* are known

$$Z = \sqrt{\text{Resistance}^2 + \text{Reactance}^2} = \sqrt{R^2 + X^2}$$

Determine the Impedance of an ac circuit when the resistance is 6 ohms, and the reactance is 8 ohms.

$$Z = \sqrt{R^2 + X^2} = \sqrt{36 + 64} = \sqrt{100} = 10 \text{ Ohms}$$

C. When *resistance, Inductive reactance,* and *capacitive reactance* are known:

$$Z = \sqrt{R^2 + (X_L - X_C)^2}$$

Determine the Impedance of an ac circuit that has a resistance of 6 ohms, an Inductive reactance of 18 ohms, and a capacitive reactance of 10 ohms.

$$Z = \sqrt{R^2 + (X_L - X_C)^2}$$

$$Z = \sqrt{6^2 + (18 - 10)^2} = \sqrt{6^2 + (8)^2}$$

$$= \sqrt{36 + 64} = \sqrt{100} = 10 \text{ Ohms}$$

TO FIND REACTANCE

Reactance (X)

Reactance in a circuit is the opposition to an alternating current caused by inductance and capacitance, equal to the difference between capacitive and inductive reactance. Reactance is Expressed in ohms.

A. Inductive Reactance X_L

Inductive reactance is that element of reactance in a circuit caused by self-inductance.

$$X_L = 2 \times 3.1416 \times \text{Frequency} \times \text{Inductance}$$
$$= 6.28 \qquad X \qquad F \qquad X \qquad L$$

Determine the reactance of a 4-Henry coil on a 60-cycle ac circuit.

$$X_L = 6.28 \times F \times L = 6.28 \times 60 \times 4 = 1507 \text{ Ohms}$$

B. Capacitive Reactance X_C

Capacitive reactance is that element of reactance in a circuit caused by capacitance.

$$X_C = \frac{1}{2 \times 3.1416 \times \text{Frequency} \times \text{Capacitance}}$$

$$= \frac{1}{6.28 \qquad X \qquad F \qquad X \qquad C}$$

Determine the reactance of a 4 microfarad condenser on a 60-cycle ac circuit

$$X_C = \frac{1}{6.28 \times F \times C} = \frac{1}{6.28 \times 60 \times 0.000004}$$

$$= \frac{1}{0.0015072} = 663 \text{ Ohms}$$

A Henry is a unit of inductance equal to the inductance of a circuit in which the variation of a current at the rate of 1 ampere per second induces an electromotive force of 1 volt.

FULL-LOAD CURRENT IN AMPERES: DIRECT-CURRENT MOTORS

HP	Armature Voltage Rating*					
	90 V	120 V	180 V	240 V	500 V	550 V
¼	4.0	3.1	2.0	1.6	–	–
1/3	5.2	4.1	2.6	2.0	–	–
½	6.8	5.4	3.4	2.7	–	–
¾	9.6	7.6	4.8	3.8	–	–
1	12.2	9.5	6.1	4.7	–	–
1½	–	13.2	8.3	6.6	–	–
2	–	17	10.8	8.5	–	–
3	–	25	16	12.2	–	–
5	–	40	27	20	–	–
7½	–	58	–	29	13.6	12.2
10	–	76	–	38	18	16
15	–	–	–	55	27	24
20	–	–	–	72	34	31
25	–	–	–	89	43	38
30	–	–	–	106	51	46
40	–	–	–	140	67	61
50	–	–	–	173	83	75
60	–	–	–	206	99	90
75	–	–	–	255	123	111
100	–	–	–	341	164	148
125	–	–	–	425	205	185
150	–	–	–	506	246	222
200	–	–	–	675	330	294

These values of full-load currents* are for motors running at base speed.

*These are average dc quantities.

Reprinted with permission from NFPA 70®, *National Electrical Cotie*®, 2023 edition, Table 430.247, Copyright © 2022, National Fire Protection Association, Quincy, MA 02169. This reprinted material is not the complete and official position of the NFPA or the referenced subject, which is represented only by the standard in its entirety.

DIRECT-CURRENT MOTORS

Terminal Markings

Terminal markings are used to tag terminals to which connections are to be made from outside circuits.

Facing the end opposite the drive (commutator end), the standard direction of shaft rotation is counterclockwise.

A_1 and A_2 indicate armature leads.
S_1 and S_2 indicate series-field leads.
F_1 and F_2 indicate shunt-field leads.

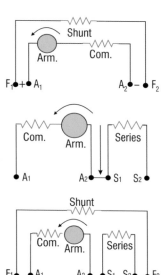

Shunt-Wound Motors

To change rotation, reverse either armature leads or shunt leads. **Do not** reverse both armature and shunt leads

Series-Wound Motors

To change rotation, reverse either armature leads or series leads. **Do not** reverse both armature and series leads

Compound-Wound Motors

To change rotation, reverse either armature leads or both the series and shunt leads. **Do not** reverse all three sets of leads.

Note: Standard rotation for a dcgenerator is clockwise.

FULL-LOAD CURRENT IN AMPERES: SINGLE-PHASE ALTERNATING-CURRENT MOTORS

HP	115 V	200 V	208 V	230 V
⅙	4.4	2.5	2.4	2.2
¼	5.8	3.3	3.2	2.9
1/3	7.2	4.1	4.0	3.6
½	9.8	5.6	5.4	4.9
¾	13.8	7.9	7.6	6.9
1	16	9.2	8.8	8.0
1½	20	11.5	11	10
2	24	13.8	13.2	12
3	34	19.6	18.7	17
5	56	32.2	30.8	28
7½	80	46	44	40
10	100	57.5	55	50

The voltages listed are rated motor voltages. The currents listed shall be permitted for system voltage ranges of 110 to 120 and 220 to 240 volts.

Source: NFPA 70®, *National Electrical Code®,* 2023 edition, NFPA, Quincy, MA, 2022, Table 430.248.

SINGLE-PHASE MOTOR USING STANDARD THREE-PHASE STARTER

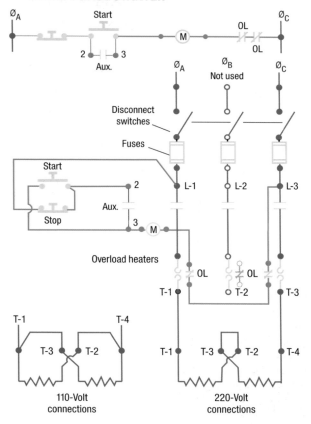

110-Volt connections

220-Volt connections

(M) = Motor starter coil

⏻ SINGLE-PHASE MOTORS

Split-Phase, Squinel-Cage, Dual-Voltage Motor

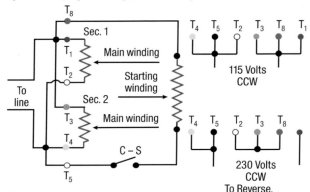

115 Volts
CCW

230 Volts
CCW
To Reverse,
Interchange 5 and 8

Classes of Single-Phase Motors

1. Split-Phase
   A. Capacitor Start
   B. Repulsion Start
   C. Resistance Start
   D. Split Capacitor

2. Commutator
   A. Repulsion
   B. Series

Terminal Color Marking

T_1 Blue •	T_3 Orange •	T_5 Black •
T_2 White	T_4 Yellow	T_8 Red •

Note: Split-phase motors are usualy fractional horsepower. The majority of electric motors used in washing machines, refrigerators, etc. are of the split-phase type.

To change the speed of a split-phase motor, the number of poles must be changed.

1. Addition of running winding
2. Two starting windings and two running windings
3. Consequent pole connections

SINGLE-PHASE MOTOR

Split-Phase, Squinel-Cage, Motor

A. Resistance Start:

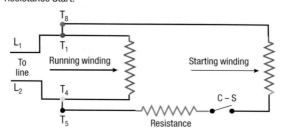

Centrifugal switch o (cs) opens after reaching 75% of normal speed.

B. Capacitor Start:

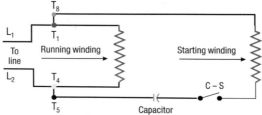

Note: 1. A resistance start motor has a resistance connected in series with the starting winding.

2. The capacitor start motor is employed where a high starting torque is required.

RUNNING OVERLOAD UNITS

Kind of Motor	Supply System	Number and Location of Overload Units Such as Trip Coils or Relays
1-phase ac or dc	2-wire, 1-phase ac or dc, ungrounded	1 in either conductor
1-phase ac or dc	2-wire, 1-phase ac or dc, one conductor ungrounded	1 in ungrounded conductor
1-phase ac or dc	3-wire, 1-phase ac or dc, grounded neutral conductor	1 in either ungrounded conductor
1-phase ac	Any 3-phase	1 in ungrounded conductor
2-phase ac	3-wire, 2-phase ac, ungrounded	2, one in each phase
2-phase ac	3-wire, 2-phase ac, one conductor grounded	2 in ungrounded conductors
2-phase ac	4-wire, 2-phase ac, grounded or ungrounded	2, one for each phase in ungrounded conductors
2-phase ac	Grounded neutral or 5-wire, 2-phase ac, ungrounded	2, one for each phase in any ungrounded phase wire
3-phase ac	Any 3-phase	3, one in each phase*

*Exception: An overload unit in each phase shall not be required where overload protection is provided by other approved means.
Source: NFPA 70®, National Electrical Code®, NFPA, 2023 edition, Quincy, MA, 2022, Table 430.37.

 MOTOR BRANCH-CIRCUIT PROTECTIVE DEVICES: MAXIMUM RATING OR SETTING

	Percent of Full-Load Current			
Type of Motor	Nontime Delay Fuse	Dual Element (Time-Delay) Fuse[1]	Instantaneous Trip Breaker	Inverse Time Breaker[2]
Single-phase motors	300	175	800	250
AC polyphase motors other than wound rotor	300	175	800	250
Squirrel cage— other than Design B energy-efficient— and Design B premium-efficiency	300	175	800	250
Design B energy-efficient and Design B premium-efficiency	300	175	1100	250
Synchronous[3]	300	175	800	250
Wound rotor	150	150	800	150
Direct current (constant voltage)	150	150	250	150

For certain exceptions to the values specified, see 430.54.
1 The values in the Nontime Delay Fuse column apply to Time-Delay Class CC fuses.
2 The values given in the last column also cover the ratings of nonadjustable inverse lime types of circuit breakers that can be modified as in 430.52(C)(1)(a) and (C)(1)(b).
3 Synchronous motors of the low-torque, low-speed type (usually 450 rpm or lower), such as those used to drive reciprocating compressors, pumps, and so forth, that start unloaded, do not require a fuse rating or circuit-breaker setting in excess of 200% of ful-load current.
Source: NFPA 70®, National Electrical Code®, 2023 edition, NFPA, Quincy, MA, 2022, Table 430.52(C)(1).
Note: Where the result of the calculation for the branch circuit protective device does not correspond with a standard size fuse or circuit breaker, see 430.52(C)(a).
Note: Where the rating specified in Table 430.52(C)(1), or the rating modified by 430.52(C)(1)(a), is not sufficient for the starting current of the motor, see 430.52(C)(b).

FULL-LOAD CURRENT: THREE-PHASE ALTERNATING-CURRENT MOTORS

HP	Induction-Type Squirrel Cage and Wound Rotor (Amperes)							Synchronous-Type Unity Power Factor* (Amperes)			
	115 Volts	200 Volts	208 Volts	230 Volts	460 Volts	575 Volts	2300 Volts	230 Volts	460 Volts	575 Volts	2300 Volts
½	4.4	2.5	2.4	2.2	1.1	0.9	-	-	-	-	-
¾	6.4	3.7	3.5	3.2	1.6	1.3	-	-	-	-	-
1	8.4	4.8	4.6	4.2	2.1	1.7	-	-	-	-	-
1½	12.0	6.9	6.6	6.0	3.0	2.4	-	-	-	-	-
2	13.6	7.8	7.5	6.8	3.4	2.7	-	-	-	-	-
3	-	11.0	10.6	9.6	4.8	3.9	-	-	-	-	-
5	-	17.5	16.7	15.2	7.6	6.1	-	-	-	-	-
7½	-	25.3	24.2	22	11	9	-	-	-	-	-
10	-	32.2	30.8	28	14	11	-	-	-	-	-
15	-	48.3	46.2	42	21	17	-	-	-	-	-
20	-	62.1	59.4	54	27	22	-	-	-	-	-
25	-	78.2	74.8	68	34	27	-	53	26	21	-
30	-	92	88	80	40	32	-	63	32	26	-
40	-	120	114	104	52	41	-	83	41	33	-
50	-	150	143	130	65	52	-	104	52	42	-
60	-	177	169	154	77	62	16	123	61	49	12
75	-	221	211	192	96	77	20	155	78	62	15
100	-	285	273	248	124	99	26	202	101	81	20
125	-	359	343	312	156	125	31	253	126	101	25
150	-	414	396	360	180	144	37	302	151	121	30
200	-	552	528	480	240	192	49	400	201	161	40
250	-	-	-	-	302	242	60	-	-	-	-
300	-	-	-	-	361	289	72	-	-	-	-
350	-	-	-	-	414	336	83	-	-	-	-
400	-	-	-	-	477	382	95	-	-	-	-
450	-	-	-	-	515	412	103	-	-	-	-
500	-	-	-	-	590	472	118	-	-	-	-

The voltages listed are rated motor voltages. The currents listed shall be permitted for system voltage ranges of 110 to 120, 220 to 240, 440 to 480, 550 to 600, and 2300 to 2400 volts.

*For 90% and 80% power factor, the figures shall be multiplied by 1.1 and 1.25, respectively.

Source: NFPA 70®, *National Electrical Code®,* NFPA, 2023 edition, Quincy, MA, 2022, Table 430.250.

**FULL LOAD CURRENT AND OTHER DATA:
THREE PHASE AC MOTORS**

Motor Horsepower		Motor Ampere	Size Breaker *	Size Starter	Heater Ampere **	Size Wire	Size Conduit ***
½	230 V	2.2	15	00	2.530	12	¾"
	460	1.1	15	00	1.265	12	¾
¾	230	3.2	15	00	3.680	12	¾
	460	1.6	15	00	1,840	12	¾
1	230	4.2	15	00	4.830	12	¾
	460	2.1	15	00	2.415	12	¾
1½	230	6.0	15	00	6.900	12	¾
	460	3.0	15	00	3.450	12	¾
2	230	6.8	15	0	7.820	12	¾
	460	3.4	15	00	3.910	1?	¾
3	230	9.6	20	0	11.040	12	¾
	460	4.8	15	0	5,520	12	¾
5	230	15.2	30	1	17.480	12	¾
	460	7,6	15	0	8.740	12	¾
7½	230	22	45	1	25.300	10	¾
	460	11	20	1	12.650	12	¾
10	230	28	60	2	32.200	10	¾
	460	14	30	1	16.100	12	¾
15	230	42	70	2	48.300	6	1
	460	21	40	2	24.150	10	¾
20	230	54	100	3	62.100	4	1
	460	27	50	2	31.050	10	¾
25	230	68	100	3	78.200	4	1½
	460	34	50	2	39.100	8	1
30	230	80	125	3	92.000	3	1½
	460	40	70	3	46.000	8	1
40	230	104	175	4	119.600	1	1½
	460	52	100	3	59.800	6	1
50	230	130	200	4	149.500	00	2
	460	65	150	3	74.750	4	1½
60	230	154	250	5	177.10	000	2
	460	77	200	4	88,55	3	1½

(continued on next page

40

FULL-LOAD CURRENT AND OTHER DATA: THREE PHASE AC MOTORS

Motor Horsepower		Motor Ampere	Size Breaker	Size Starter	Heater Ampere	Size Wire	Size Conduit
75	230 V	192	300	5	220.80	250 kcmil	2½"
	460	96	200	4	110.40	1	1½
100	230	248	400	5	285.20	350 kcmil	3
	460	124	200	4	142.60	2/0	2
125	230	312	500	6	358.80	600kcmil	3½
	460	156	250	5	179.40	000	2
150	230	360	600	6	414.00	700 kcmil	4
	460	180	300	5	207.00	0000	2½

* Overcurrent device may have to be increaaed due to starting current and load conditions
 See NEC 430,52, Table 430.52. Wire size based on 75°C (187°F) terminations and 75°C (157°F) insulation,
** Overload heater must be based on motor nameplate and sized per NEC 430.32.
***Conduit size based on rigid metal conduit with some spa re capacity, For minimum size and other conduit types, see *NEC* Appendix C, or *Ugly's* pages 83-103.

 MOTOR AND MOTOR CIRCUIT CONDUCTOR PROTECTION

Motors can have large starting currents three to five times or more than that of the actual motor current. In order for motors to start, the motor and motor circuit conductors are allowed to be protected by circuit breakers and fuses at values that are higher than the actual motor and conductor ampere ratings. These larger overcurrent devices do not provide overload protection and will only open upon short circuits or ground faults. Overload protection must be used to protect the motor based on the actual nameplate amperes of the motor. This protection is usually in the form of heating elements in manual or magnetic motor starters. Small motors such as waste disposal motors have a red reset button built into the motor

General Motor Rules

- Use full-load current from Tables instead of nameplate.
- Branch Circuit Conductors: Use 125% of full-load current to find conductor size.
- Branch Circuit OCP Size: Use percentages given in Tables for full- load current, (*Ugly's* pages 32) 34, and 39)
- Feeder Conductor Size: 125% of largest motor and sum of the rest.
- Feeder OCP: Use largest OCP plus rest of full-load currents.

(See examples on Ugly's page 42.)

MOTOR BRANCH CIRCUIT AND FEEDER EXAMPLE

General Motor Applications

Branch circuit conductors:

Use full-load, three-phase currents, from the table on *Ugly's* page 39 or 2023 *NEC* Table 430.250, 50-HP, 480-volt, 3-phase, motor design B, 75-degree terminations = 65 Amps
125% of full-load current [*NEC* 430.22] (*Ugly's* page 41)
125% of 65 Amps =**81.25 Amps** conductor selection ampacity

Branch circuit overcurrent device: *NEC* 430.52(C)(1)

(Branch circuit short-circuit and ground fault protection)
Use percentages given in *Ugly's* page 38 or *NEC* Table 430.52(C)(1) for **Type** of circuit breaker or fuse used.
50-HP ,480-volt, 3-phase motor =65 Amps (*Ugly's* page 39)
Nontime fuse=300% (*Ugly's* page 38)
300% of 65 Amps=195 Amps. *NEC* 430.52(C)(1)(a) Next size allowed *NEC* Table 240.6(A)=**200-amp fuse.**

Feeder connectors:

For 50-HP and 30-HP, 480-volt, 3-phase, design B motors on same feeder:
Use 125% of the largest full-load current and 100% of the rest. (NEC 430.24)
50-HP ,480-volt, 3-phase motor = 65 Amps; 30-HP, 480-volt, 3-phase motor = 40 Amps (125% of 65 Amps) +40 Amps =**121.25 Amps** conductor selection ampacity

Feeder overcurrent device: *NEC* 430.62(A) (specific load)

(Feeder short-circuit and ground-fault protection)
Use largest overcurrent protection device **plus** full-load currents of the rest of the motors.
50 HP = 200 Amp fuse   (65 FLC)
30 HP = 125 Amp fuse   (40 FLC)
200 Amp fuse + 40 Amp (FLC) = 240 Amp. Do not exceed this value on feeder. The next standard size fuse below 240 is a **225-amp** fuse.

LOCKED-ROTOR INDICATING CODE LETTERS

Code Letter	Kilovolt-Ampere/ Horsepower with Locked Rotor	Code Letter	Kilovolt-Ampere/ Horsepower with Locked Rotor
A	0-3.14	L	9.0-9.99
B	3.15-3.54	M	10.0-11.19
C	3.55-3.99	N	11.2-12.49
D	4.0-4.49	P	12.5-13.99
E	4.5-4.99	R	14.0-15.99
F	5.0-5.59	S	16.0-17.99
G	5.6-8.29	T	18.0-19.99
H	6.3-7.09	U	20.0-22.39
J	7.1-7.99	V	22.4 and up
K	8.0-8.99		

Source: NFPA 70®, National Electrical Code®, NFPA, 2023 edition, Quincy, MA, 2022, Table 430.7(B), as modified.

The *National Electrical Code®* requires that all alternating-current motors rated ½ HP or more (except for polyphase wound-rotor motors) must have code letters on their nameplates indicating motor input with locked rotor (in kilovolt-amperes per horsepower). The motor's horsepower, voltage, and locked-rotor code letter are needed to calculated the motor's locked-rotor current. Use the folowing formulas:

Single-Phase Motors:

$$\text{Locked-Ftotor Current} = \frac{HP \times KVA_{hp} \times 1000}{E}$$

Three-Phase Motors:

$$\text{Locked-Ftotor Current} = \frac{HP \times KVA_{hp} \times 1000}{E \times 1.73}$$

Example: What is the maximum locked-rotor current for a 480-volt, 25-HP, code letter F motor?
(From the above table, code letter F = 5.59 KVA$_{hp}$)

$$I = \frac{HP \times KVA_{hp} \times 1000}{E \times 1.73} = \frac{25 \times 5.59 \times 1000}{480 \times 1.73} = \textbf{168.29 Amps}$$

MAXIMUM MOTOR LOCKED- ROTOR CURRENT IN AMPERES. SINGLE PHASE

HP	115 V	208 V	230 V	HP	115 V	208 V	230 V
½	58.8	32.5	29.4	3	204	113	102
¾	82.8	45.8	41.4	5	336	186	168
1	96	53	48	7½	480	265	240
1½	120	66	60	10	1000	332	300
2	144	80	72				

Note: For use only with 430.110, 440.12, 440.41, and 455.8(C).
Source: NFPA 70. National Electrical Code®. NFPA. Quincy, MA. 2020. Table 430.251(A). as modified.

MAXIMUM MOTOR LOCKED ROTOR CURRENT INAMPERES,TWO AND THREE PHASE. DESIGN B, C, AND D*

HP	115 V	200 V	208 V	230 V	460 V	575 V
½	40	23	22.1	20	10	8
¾	50	28.8	27.6	25	12.5	10
1	60	34.5	33	30	15	12
1½	80	46	44	40	20	16
2	100	57.5	55	50	25	20
3	–	73.6	71	64	32	25.6
5	–	105.8	102	92	46	36.8
7½	–	146	140	127	63.5	50.8
10	–	186.3	179	162	81	64.8
15	–	267	257	232	116	93
20	–	334	321	290	145	116
25	–	420	404	365	183	146
30	–	500	481	435	218	174
40	–	667	641	580	290	232
50	–	834	802	725	363	290
60	–	1001	962	870	435	348
75	–	1248	1200	1085	543	434
100	–	1668	1603	1450	725	580
125	–	2087	2007	1815	908	726
150	–	2496	2400	2170	1085	868
200	–	3335	3207	2900	1450	1160

* Design A motors are not limited to a maximum starling current or kicked-rotor current.
Note: For use only with 430.110, 440.12, 440.41, and 455.8(C).
Source: NFPA 70®. National Electrical Code®. 2023 edition, NFPA, Quincy, MA. 2022, Table 430.251(B). as modified.

THREE-PHASE AC MOTOR WINDINGS AND CONNECTIONS

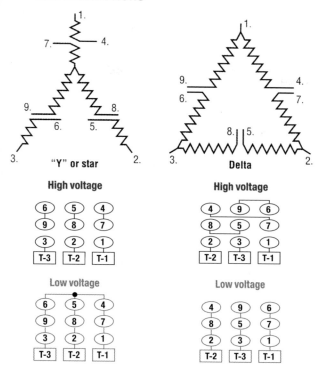

"Y" or star

Delta

High voltage

6	5	4
9	8	7
3	2	1
T-3	T-2	T-1

High voltage

4	9	6
8	5	7
2	3	1
T-2	T-3	T-1

Low voltage

6	5	4
9	8	7
3	2	1
T-3	T-2	T-1

Low voltage

4	9	6
8	5	7
2	3	1
T-2	T-3	T-1

Note: 1. The most important part of any motor is the nameplate. check the data given on the plate before making the connections.
2. To change rotation direction of 3-phase motor, swap any two T-leads.

THREE-WIRE STOP-START STATION

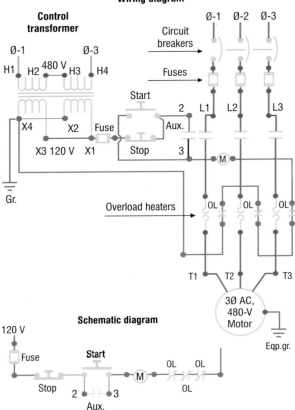

Wiring diagram

Control transformer

Ø-1   Ø-3
H1   H2 480 V H3   H4

Ø-1   Ø-2   Ø-3

Circuit breakers

Fuses

X4   X2   Fuse   Start
X3 120 V X1

L1   L2   L3

2
Aux.
Stop
3

M

Gr.

Overload heaters

OL   OL   OL

T1   T2   T3

3Ø AC,
480-V
Motor

Eqp.gr.

Schematic diagram

120 V

Fuse

Start

Stop   2   3
Aux.

M   OL   OL
OL

M = Motor starter coil

Note: Controls and motor are of dirrerent voltages.

46

⚡ TWO THREE-WIRE STOP-START STATIONS

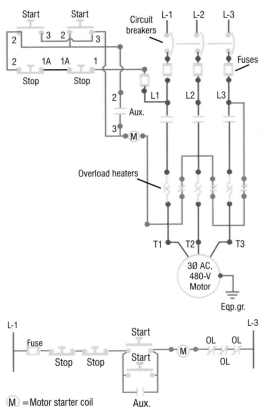

(M) = Motor starter coil

Note: Controls and motor are of dirrerent voltages.
If low-voltage controls are used, see *Ugly's* page 46 for control transformer connections.

Wiring diagram

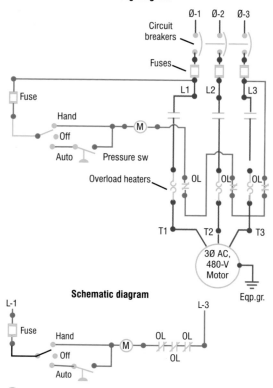

Schematic diagram

(M) = Motor starter coil

Note: Controls and motor are of the same voltage.
If low-voltage controls are used, see *Ugly's* page 46 for control transformer connections.

48

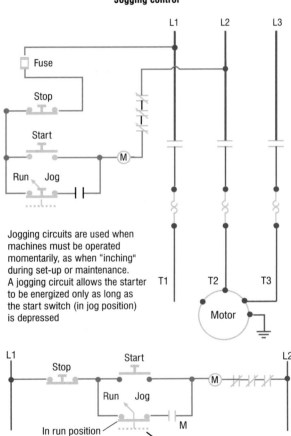

JOGGING WITH SELECTOR SWITCH

Jogging control

Fuse

Stop

Start

(M)

Run   Jog

Jogging circuits are used when
machines must be operated
momentarily, as when "inching"
during set-up or maintenance.
A jogging circuit allows the starter
to be energized only as long as
the start switch (in jog position)
is depressed

L1   L2   L3

T1   T2   T3

Motor

L1   Stop   Start   L2
(M)

Run   Jog

M

In run position

in jog position

(M) = Motor starter coil

49

VOLTAGE DROP CALCULATIONS: INDUCTANCE NEGLIGIBLE

Vd = Voltage Drop
I = Current in Conductor (Amperes)
L = One-way Length of Circuit (Feet)
Cm = Cross Sectional Area of Conductor (Circular Mils) (page 71)
K = Resistance in Ohms of 1 Circular Mil Foot of Conductor
  K = 12.9 for Copper Conductors @ 75°C (167°F)
  K = 21.2 for Aluminum Conductors @ 75°C (167°F)

Note: K value changes with temperature and other factors. See *NEC* Chapter 9, Table 8, Note 1.

Single-Phase Circuits

$$Vd = \frac{2K \times L \times I}{Cm} \quad \text{or} \quad {}^*Cm = \frac{2K \times L \times I}{Vd}$$

Three-Phase Circuits

$$Vd = \frac{1.73K \times L \times I}{Cm} \quad \text{or} \quad {}^*Cm = \frac{1.73K \times L \times I}{Vd}$$

* *Note:* Always check ampacity tables to ensure conductor's ampacity is equal to load <u>after voltage drop calculation.</u>

Refer to *Ugly's* pages 71-82 for conductor size, type, and ampacity,
See *Ugly's* pages 51-52 for examples

 VOLTAGE DROP EXAMPLES

Distance (One Way) for 2% Voltage Drop for 120 Volts Single Phase

AMPS	VOLTS	12 AWG	10 AWG	8 AWG	6 AWG	4 AWG	3 AWG	2 AWG	1 AWG	1/0 AWG
20	120	30	48	77	122	194	245	309	389	491
	240	60	96	154	244	388	490	618	778	982
30	120		32	51	81	129	163	206	260	327
	240		64	102	162	258	326	412	520	654
40	120			38	61	97	122	154	195	246
	240			76	122	194	244	308	390	492
50	120				49	78	96	123	156	196
	240				98	156	196	246	312	392
60	120					65	82	103	130	164
60	240					190	164	206	260	328
70	240					111	140	176	222	281
80	240						122	154	195	246
90	240							137	173	218
100	240								156	196

(See Footnotes on Page 51 Concerning Circuit Load Limitations.)

VOLTAGE DROP EXAMPLES

Typical Voltage Drop Values Based on Conductor Size and One-Way Length* (60°C [140°F] Termination and Insulation)

25 Feet							
	12 AWG	10 AWG	8 AWG	6 AWG	4 AWG	3 AWG	2 AWG	1 AWG
20 A	1.98	1.24	0.78	0.49	0.31	0.25	0.19	0.15
30		1.86	1.17	0.74	0.46	0.37	0.29	0.23
40			1.56	0.98	0.62	0.49	0.39	0.31
50				1.23	0.77	0.61	0.49	0.39
60					0.93	0.74	0.58	0.46

50 Feet							
	12 AWG	10 AWG	8 AWG	6 AWG	4 AWG	3 AWG	2 AWG	1 AWG
20 A	3.95	2.49	1.56	0.98	0.62	0.49	0.39	0.31
30		3.73	2.34	1.47	0.93	0.74	0.58	0.46
40			3.13	1.97	1.24	0.98	0.78	0.62
50				2.46	1.55	1.23	0.97	0.77
60					1.85	1.47	1.17	0.92

75 Feet							
	12 AWG	10 AWG	8 AWG	6 AWG	4 AWG	3 AWG	2 AWG	1 AWG
20 A	5.93	3.73	2.34	1.47	0.93	0.74	0.58	0.46
30		5.59	3.52	2.21	1.39	1.10	0.87	0.69
40			4.69	2.95	1.85	1.47	1.17	0.92
50				3.69	2.32	1.84	1.46	1.16
60					2.78	2.21	1.75	1.39

100 Feet							
	12 AWG	10 AWG	8 AWG	6 AWG	4 AWG	3 AWG	2 AWG	1 AWG
20 A	7.90	4.97	3.13	1.97	1.24	0.98	0.78	0.62
30		7.46	4.69	2.95	1.85	1.47	1.17	0.92
40			6.25	3.93	2.47	1.96	1.56	1.23
50				4.92	3.09	2.45	1.94	1.54
60					3.71	2.94	2.33	1.85

125 Feet							
	12 AWG	10 AWG	8 AWG	6 AWG	4 AWG	3 AWG	2 AWG	1 AWG
20 A	9.88	6.21	3.91	2.46	1.55	1.23	0.97	0.77
30		9.32	5.86	3.69	2.32	1.84	1.46	1.16
40			7.81	4.92	3.09	2.45	1.94	1.54
50				6.15	3.86	3.06	2.43	1.93
60					4.64	3.68	2.92	2.31

150 Feet							
	12 AWG	10 AWG	8 AWG	6 AWG	4 AWG	3 AWG	2 AWG	1 AWG
20 A	11.85	7.46	4.69	2.95	1.85	1.47	1.17	0.92
30		11.18	7.03	4.42	2.78	2.21	1.75	1.39
40			9.38	5.90	3.71	2.94	2.33	1.85
50				7.37	4.64	3.68	2.92	2.31
60					5.56	4.41	3.50	2.77

A 2-wire, 20-amp circuit using 12 AWG with a one-way distance ot 25 feet will drop 1.98 volts.
120 Volts - 1.98 Volts = 118,02 Volts as the load voltage.
240 Volts - 1.98 Volts = 238.02 Volts as the load voltage.

*Bettor economy and efficiency will result using the voltage drop method on page 50.
A continuous load cannot exceed 80% ol the circuit rating,
A motor or heating load cannot exceed 80% ot the circuit rating.
For motor overcurrent devices and conductor sizing, see Ugly's pages 40-42,

VOLTAGE DROP CALCULATION EXAMPLES

Single-Phase Voltage Drop

What is the voltage drop of a 240-volt, single-phase circuit consisting of #8 THWN copper conductors feeding a 30-ampere load that is 150 feet in length?

Voltage Drop Formula (see *Ugly's* page 50)

$$V_d = \frac{2K \times L \times I}{Cm} = \frac{2 \times 12.9 \times 150 \times 30}{16510} = \frac{116100}{16510} = \textbf{7 Volts}$$

Percentage voltage drop = 7 Volts/240 Volts = 0.029 = **2.9%**
Voltage at load = 240 Volts − 7 Volts = **233 Volts**

Three-Phase Voltage Drop

What is the voltage drop of a 480-volt, 3-phase circuit consisting of 250-kcmil THWN copper conductors that supply a 250-ampre load that is 500 feet from the source?

250 kcmil = 250000 circular mils

Voltage Drop Formula (see *Ugly's* page 50)

$$V_d = \frac{1.73K \times L \times I}{Cm} = \frac{1.73 \times 12.9 \times 500 \times 250}{250000} = \frac{2789625}{250000} = \textbf{11 Volts}$$

Percentage voltage drop = 11 Volts/480 Volts =0.0229 = **2.29%**
Voltage at load = **480** Volts - 11 Volts = **469 Volts**

Note: Always check ampacity tables for conductors selected.

Refer to *Ugly's* pages 71-82 for conductor size, type, and ampacity.

⚡ SHORT-CIRCUIT CALCULATION

(Courtesy of Cooper Bussmann)

Basic Short-Circuit Calculation Procedure

1. Determine transformer full-load amperes frdm either:
   a) Nameplate
   b) Formula:

$$3\emptyset \text{ transformer } \quad I_{l.l.} = \frac{KVA \times 1000}{E_{L-L} \times 1.732}$$

$$1\emptyset \text{ transformer } \quad I_{l.l.} = \frac{KVA \times 1000}{E_{L-L}}$$

2. Find transformer multiplier.

$$\text{Multiplier} = \frac{100}{*\%Z_{trans}}$$

3. Determine transformer let-through short-circuit Current.**

$$I_{s.c.} = I_{l.l.} \times \text{Multiplier}$$

4. Calculate "f" factor.

$$3\emptyset \text{ faults} \quad f = \frac{1.732 \times L \times I_{3\emptyset}}{C \times E_{L-L}}$$

$$\begin{array}{l}\textbf{1}\emptyset \textbf{ line-to-line (L-L) faults} \\ \text{on } 1\emptyset \text{ Center Tapped} \\ \text{Transformer}\end{array} \quad f = \frac{2 \times L \times I_{L-L}}{C \times E_{L-L}}$$

$$\begin{array}{l}\textbf{1}\emptyset \textbf{ line-to-neutral (L-N)} \\ \textbf{faults on } 1\emptyset \text{ Center Tapped} \\ \text{Transformer}\end{array} \quad f = \frac{2 \times L \times I_{L-N}***}{C \times E_{L-N}}$$

L = Length (feet) of Conductor to the fault
C = Constant from table C (page 551 for conductors and busway. For parallel runs, multiply C values by the number of conductors per phase.
I = Available short-circuit current in amperes at beginning of circuit.

5. Calculate "M" (multiplier)

$$M = \frac{1}{1 + f}$$

6. Calculate the available short-circuit symmetrical RMS current at the point of fault.

$$I_{s.c. \text{ sym RMS}} = I_{s.c.} \times M$$

⚡ SHORT-CIRCUIT CALCULATION

(Courtesy of Cooper Bussmann)

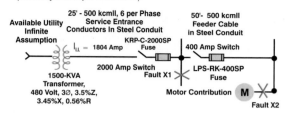

Example: Short-Circuit Calculation

(Fault #1)

1. $I_{L.L.} = \dfrac{KVA \times 1000}{E_{L-L} \times 1.732} = \dfrac{1500 \times 1000}{480 \times 1.732} = \textbf{1804 Amps}$

2. $\text{Multiplier} = \dfrac{100}{\ast \, \%Z_{trans}} = \dfrac{100}{3.5} = \textbf{28.57}$

3. $I_{s.c.} = 1804 \times 28.57 = \textbf{51540 A}$

4. $f = \dfrac{1.732 \times L \times I_{3\emptyset}}{C \times E_{L-L}} = \dfrac{1.73 \times 25 \times 51540}{6 \times 22185 \times 480} = \textbf{0.0349}$

5. $M = \dfrac{1}{1 + f} = \dfrac{1}{1 + 0.0349} = \textbf{0.9663}$

6. $I_{s.c. \, sym \, RMS} = I_{s.c.} \times M = 51540 \times 0.9663 = \textbf{49803 Amps}$

$I_{s.c. \, motor \, contrib} = 4 \times 1804 = \textbf{7216 Amps}$

$I_{total \, s.c. \, sym \, RMS} = 49803 + 7216 = \textbf{57019 Amps}$

(Fault #2)

4. Use $I_{s.c. \, sym \, RMS}$ @ Fault X_1 to calculate "f"

$f = \dfrac{1.73 \times 50 \times 49803}{22185 \times 480} = \textbf{0.4050}$

5. $M = \dfrac{1}{1 + 0.4050} = \textbf{0.7117}$

6. $I_{s.c. \, sym \, RMS} = 49803 \times 0.7117 = \textbf{35445 Amps}$

$I_{sym \, motor \, contrib} = 4 \times 1804 = \textbf{7216 Amps}$

$I_{total \, s.c. \, sym \, RMS} = 35445 + 7216 = \textbf{42661 Amps}$

SHORT-CIRCUIT CALCULATION (Courtesy of Cooper Bussmann)

"C" Values for Conductors

AWG or MCM	Copper Three Single Conductors						Copper Three Conductor Cable					
	Steel Conduit			Nonmagnetic Conduit			Steel Conduit			Nonmagnetic Conduit		
	600 V	5 KV	15 KV	600 V	5 KV	15 KV	600 V	5 KV	15 KV	600 V	5 KV	15 KV
12	617	617	617	617	617	617	617	617	617	617	617	617
10	981	981	981	981	981	981	981	981	981	981	981	981
8	1557	1551	1557	1558	1555	1558	1559	1557	1559	1559	1558	1559
6	2425	2406	2389	2430	2417	2406	2431	2424	2414	2433	2428	2420
4	3806	3750	3695	3825	3789	3752	3830	3811	3778	3837	3823	3798
3	4760	4760	4760	4802	4802	4802	4760	4790	4760	4802	4802	4802
2	5906	5736	5574	6044	5926	5809	5989	5929	5827	6087	6022	5957
1	7292	7029	6758	7493	7306	7108	7454	7364	7188	7579	7507	7364
1/0	8924	8543	7973	9317	9033	8590	9209	9086	8707	9472	9372	9052
2/0	10755	10061	9389	11423	10877	10318	11244	11015	10500	11703	11528	11052
3/0	12843	11804	11021	13923	13048	12360	13656	13333	12613	14410	14118	13461
4/0	15082	13605	12542	16673	15351	14347	16391	15890	14813	17482	17019	16012
250	16483	14924	13643	18593	17120	15865	18310	17850	16465	19779	19352	18001
300	18176	16292	14768	20867	18975	17408	20617	20051	18318	22524	22938	20163
350	19703	17385	15678	22736	20526	18672	22646	21914	19821	24904	24126	21982
400	20565	18235	16365	24296	21786	19731	24253	23371	21042	26915	26044	23517
500	22185	19172	17492	26706	23277	21329	26980	25449	23125	30028	28712	25916
600	22965	20567	17962	28033	25203	22097	28752	27974	24896	32236	31258	27766
750	24136	21386	18888	28303	25430	22690	31050	30024	26932	32404	31338	28303
1000	25278	22539	19923	31490	28083	24887	33864	32688	29320	37197	35748	31959

Reprinted with permission. Cooper Bussmann, Inc.; www.cooperbussmann.com.

 (continued on next page)

SHORT-CIRCUIT CALCULATION

(Courtesy of Cooper Bussmann)

Notes:
*Transformer impedance (Z) helps to determine what the short circuit current will be at the transformer secondary. Transformer impedance is determined as follows:

The transformer secondary Is short circuited. Voltage is applied to the primary, which causes full-load current to flow in the secondary. This applied voltage divided by the rated primary voltage is the impedance of the transformer.

Example:

For a 480-volt rated primary, if 9.6 volts causes secondary full-load current to flow through the shorted secondary, the transformer impedance is 9.6 +480 =0.02 = 2%Z.

In addition, U.L. listed transformers 25KVA and larger have a ±10% impedance tolerance. Short-circuit amperes can be affected by this — tolerance.

**Motor short-circuit contribution, if significant, may be added to the transformer secondary short-circuit current value as determined in Step 3. Proceed with this adjusted figure through Steps 4, 5, and 6. A practical estimate of motor short-circuit contribution is to multiply the total motor current in amperes by 4.

***The L-N fault Current is higher than the L-L fault current at the secondary terminals of a single-phase center-tapped transformer. The short-circuit Current available (I) for this case in Step 4 should be adjusted at the transformer terminals as follows:

At L-N center tapped transformer terminals,

I_{L-N} = 1.5 x I_{L-L} **at Transformer Terminals.**

COMPONENT PROTECTION
(Courtesy of Cooper Bussmann)

How to Use Current-Limitation Charts

Example: An 800-amps circuit and an 800-amps, low-peak, current-limiting, time-delay fuse

Horn to Use the Let-Through Charts:

Using the example above, one can determine the pertinent let-through data for the KRP-C-800SP ampere, low-peak fuse. The let-through chart pertaining to the 800-amps, low-peak fuse is illustrated.

A. Determine the PEAK let-through CURRENT.
  1. Enter the chart on the Prospective Short-Circuit Current scale at 86000 amp and proceed vertically until the 800-amps fuse curve is intersected.
  2. Follow horizontally until the Instantaneous Peak Let-Through Current scale is intersected,
  3. Read the PEAK let-through CURRENT as 49000 amps. (If a fuse had not been used, the peak current would have been 198000 amps.)

B. Determine the APPARENT PROSPECTIVE RMS SYMMETRICAL let through CURRENT.
  1. Enter the chart on the Prospective Short-Circuit current scale at 86000 amps and proceed vertically until the 800-amps fuse curve is intersected.
  2. Follow horizontally until line A-B is intersected,
  3. Proceed vertically down to the Prospective Short-Circuit Current.
  4. Read the APPARENT PROSPECTIVE RMS SYMMETRICAL let-through CURRENT as 21000 amps. (The RMS SYMMETRICAL let-through CURRENT would be 86000 amps. if there were no fuse In the circuit.)

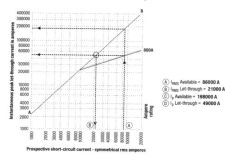

(A) I_{RMS} Available = **86000 A**
(B) I_{RMS} Let-through = **21000 A**
(C) I_p Available = **198000 A**
(D) I_p Let-through = **49000 A**

The data that can be obtained from the Fuse Let-Through Charts and their physical effects are:
1) Peak let-through current: Mechanical forces
2) Apparent prospective RMS symmetrical let-through current: Heating effect
3) Clearing time: Less than $\frac{1}{2}$ cycle when fuse is in its current-lmiting range (beyond where fuse curve intersects A– B line)

Reprinted with permission. Cooper Bussmann. Inc.; www.cooperbussmann.com.

SINGLE-PHASE TRANSFORMER CONNECTIONS

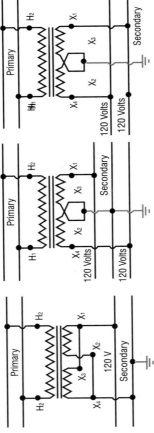

Primary — H₂ ... X₁ — **Secondary**

X₄, X₃, X₂, 120 V

Single phase to supply 120-volt lighting load. Often used lor single customer,

Primary — H₁ ... H₂ ... X₄, X₂, X₃, X₁ — **Secondary**

120 Volts / 120 Volts

Single phase to supply 120/240-volt, 3 wire lighting and power load, Used in urban distribution circuits

Primary — H₁ ... H₂ ... X₄, X₂, X₃, X₁ — **Secondary**

120 Volts / 120 Volts

Single phase for power. Used for small industrial applications.

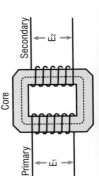

Primary E₁ — Core — Secondary E₂

A transformer is a stationary Induction device lor transferring electrical energy from one circuit to another without change of frequency. A transformer consists of two coils or windings wound upon a magnetic core of soft iron laminations and insulated from one another.

58

BUCK-AND-BOOST TRANSFORMER CONNECTIONS

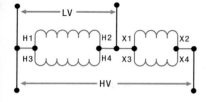

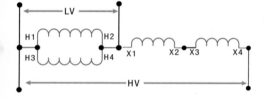

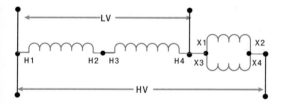

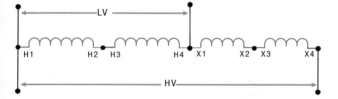

FULL-LOAD CURRENTS

	Three-Phase Transformer's Voltage (Line To Line)						Single-Phase Transformer's Voltage				
KVA Rating	208	240	480	2400	4160	KVA Rating	120	208	240	480	2400
3	8.3	7.2	3.6	.72	.416	1	8.33	4.81	4.17	2.08	.42
6	16.7	14.4	7.2	1.44	.83	3	25.0	14.4	12.5	6.25	1.25
9	25.0	21.7	10.8	2.17	1.25	5	41.7	24.0	20.8	10.4	2.08
15	41.6	36.1	18.0	3.6	2.08	7.5	62.5	36.1	31.3	15.6	3.13
30	83.3	72.2	36.1	7.2	4.16	10	83.3	48.1	41.7	20.8	4.17
45	124.9	108.3	54.1	10.8	6.25	15	125.0	72.1	62.5	31.3	6.25
75	208.2	180.4	90.2	18.0	10.4	25	208.3	120.2	104.2	52.1	10.4
100	277.6	240.6	120.3	24.1	13.9	37.5	312.5	180.3	156.3	78.1	15.6
112.5	312.5	270.6	135.3	27.1	15.6	50	416.7	240.4	208.3	104.2	20.8
150	416.4	360.9	180.4	36.1	20.8	75	625.0	360.6	312.5	156.3	31.3
225	624.6	541.3	270.6	54.1	31.2	100	833.3	480.8	416.7	208.3	41.7
300	832.7	721.7	360.9	72.2	41.6	125	1041.7	601.0	520.8	260.4	52.1
500	1387.9	1202.8	601.4	120.3	69.4	167.5	1395.8	805.3	697.9	349.0	69.8
760	2081.9	1804.3	902.1	180.4	104.1	200	1666.7	961.5	833.3	416.7	83.3
1000	2775.8	2405.7	1202.8	240.6	138.8	250	2083.3	1201.9	1041.7	520.8	104.2
1500	4163.7	3608.5	1804.3	360.9	208.2	333	2775.0	1601.0	1387.5	693.8	138.8
2000	5551.6	4811.4	2405.7	481.1	277.6	500	4166.7	2403.8	2083.3	1041.7	208.3
2500	6939.5	6014.2	3007.1	601.4	347.0						
5000	13879.0	12028.5	6014.2	1202.8	694.0						
7500	20818.5	18042.7	9021.4	1804.3	1040.9						

$$I = \frac{KVA \times 1000}{E \times 1.73} \quad \text{or} \quad KVA = \frac{E \times I \times 1.73}{1000}$$

$$I = \frac{KVA \times 1000}{E} \quad \text{or} \quad KVA = \frac{E \times I}{1000}$$

🔌 TRANSFORMER CALCULATIONS

To better understand the following formulas, review the rule of transposition in equations.

A multiplier may be removed from one side of an equation by making it a divisor on the other side; or a divisor may be removed from one side of an equation by making it a multiplier on the other side.

1. Voltage and Current: Primary (p) and Secondary (s)

Power (p) = Power (s) or $Ep \times Ip = Es \times Is$

A. $Ep = \dfrac{Es \times Is}{Ip}$

B. $Ip = \dfrac{Es \times Is}{Ep}$

C. $\dfrac{Ep \times Ip}{Es} = Is$

D. $\dfrac{Ep \times Ip}{Is} = Es$

2. Voltage and Turns in Coil:

Voltage (p) × Turns (s) = Voltage (s) × Turns (p)

or

$Ep \times Ts = Es \times Tp$

A. $Ep = \dfrac{Es \times Tp}{Ts}$

B. $Ts = \dfrac{Es \times Tp}{Ep}$

C. $\dfrac{Ep \times Ts}{Es} = Tp$

D. $\dfrac{Ep \times Ts}{Tp} = Es$

3. Amperes and Turns in Coil:

Amperes (p) × Turns (p) = Amperes (s) × Turns (s)

or

$Ip \times Tp = Is \times Ts$

A. $Ip = \dfrac{Is \times Ts}{Tp}$

B. $Tp = \dfrac{Is \times Ts}{Ip}$

C. $\dfrac{Ip \times Tp}{Is} = Ts$

D. $\dfrac{Ip \times Tp}{Ts} = Is$

SIZING TRANSFORMERS

Single-Phase Transformers

Size a 480-volt, primary, or 240/120-volt secondary, sangfe-phase transformer for the following single-phase incandescent lighting load consisting of 48 recessed fixtures each rated 2 amps. 120 volts.Each fixture has a 150-watt tamp.

*(These futures can be evenly balanced on the transformer.)

Find total volt-amperes using fixture ratings—***do not use lamp watt rating.***

> 2 Amps x 120 Volts =240 Volt-Amperes
> 240 VA x 48 =11520 VA
> 11520 VA/1000 = 11.52 KVA

The single-phase transformer that meets or exceeds this value is **15 KVA.**

*24 fighting fixtures would be connected line one to the common neutral, and 24 fighting fixtures would be connected fine two to the common neutral.

Three-Phase Transformers

Size a 480-volt, primary, or 208/120-volt, secondary. 3-phase transformer (polyphase unit) to supply one 280-volt, 3-phase, 25-KVA process heater and one 120-volt, 5-KW unit heater.

The 5-KW unit heater cannot be balanced across all three phases. The 5 KW will be on one phase only-Adding the loads directfy will undersize the transformer. Common practice is to put an imaginary load equal to the single phase load on the other two phases.
5 KW x 3 = 15 KW*
The 25 KVA is three phase; use 25 KVA.
25 KVA+15 KVA* =40 KVA
The nearest 3-phase transformer that meets or exceeds this value is a **45 KVA.**
*(KVA=KW at unity power factor) (Transformers are rated at KVA.)

SINGLE-PHASE TRANSFORMER

Primary and Secondary Amperes

A 480/240-volt, single-phase, 50-KVA transformer (Z = 2%) is to be Installed. Calculate primary and secondary amperes and short-circuit amperes.

Primary amperes:

$$I_p = \frac{KVA \times 1000}{E_p} = \frac{50 \times 1000}{480} = \frac{50000}{480} = \textbf{104 Amps}$$

Secondary amperes:

$$I_s = \frac{KVA \times 1000}{E_s} = \frac{50 \times 1000}{240} = \frac{50000}{240} = \textbf{208 Amps}$$

Short-circuit amperes:*

$$I_{sc} = \frac{I_s}{\%Z} = \frac{208}{0.02} = \textbf{10400 Amps}$$

 THREE-PHASE TRANSFORMER

Primary and Secondary Amperes

A 480/208-volt, 3-phase, 100-KVA transformer (Z = 1%) is to be installed. Calculate primary and secondary amperes and short-circuit amperes.

Primary amperes:

$$I_p = \frac{KVA \times 1000}{E_p \times 1.73} = \frac{100 \times 1000}{480 \times 1.73} = \frac{100000}{831} = \textbf{120 Amps}$$

Secondary amperes:

$$I_s = \frac{KVA \times 1000}{E_s \times 1.73} = \frac{100 \times 1000}{208 \times 1.73} = \frac{100000}{360} = \textbf{278 Amps}$$

Short-circuit amperes:*

$$I_{sc} = \frac{I_s}{\%Z} = \frac{278}{0.01} = \textbf{27800 Amps}$$

*Short-circuit amperes is the current that would flow if the transformers secondary terminals were shorted phase to phase. See *Ugly's* pages 53-56 for calculating short-circuit amperes point to point using the Cooper Bussmann method.

🔲 THREE-PHASE CONNECTIONS

Wye (Star)

Voltage from "A", "B", or "C" to Neutral = E_{PHASE} (E_P)
Voltage between A and B, B and C, or C and A = E_{LINE} (E_L)
$I_L = I_P$, if balanced.

If unbalanced,

$$I_N = \sqrt{I_A^2 + I_B^2 + I_C^2 - (I_A \times I_B) - (I_B \times I_C) - (I_C \times I_A)}$$

$$E_L = E_P \times 1.73$$

$$E_P = E_L \div 1.73$$

(True Power)
Power =

    $I_L \times E_L \times$ 1.73 x Power Factor
                (cosine)

(Apparent Power)
Volt-Amperes = $I_L \times E_L \times$ 1.73

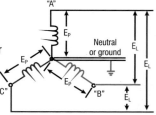

Delta

E_{LINE} (E_L) = E_{PHASE} (E_P)
$I_{LINE} = I_P \times$ 1.73
$I_{PHASE} = I_L \div$ 1.73

(True Power)
Power =

    $I_L \times E_L \times$ 1.73 x Power Factor
                (cosine)

(Apparent Power)
Volt- Amperes = $I_L \times E_L \times$ 1.73

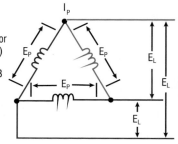

*Neutral could be ungrounded.
Also see *NEC* Article 250,
Grounding and Bonding.

64

THREE PHASE STANDARD PHASE ROTATION

Transformers

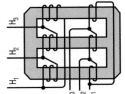

Star-Delta

Additive Polarity
30° Angular Displacement

Star-Star

Subtractive Polarity
0° Phase Displacement

Delta-Delta

Subtractive Polarity
0° Phase Displacement

TRANSFORMER CONNECTIONS

Series Connections of Low-Voltage Windings

Delta-Delta

Three-Phase Additive Polarity
High Voltage

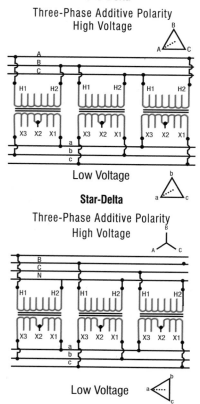

Low Voltage

Star-Delta

Three-Phase Additive Polarity
High Voltage

Low Voltage

Note: Single-phase transformers should be thoroughly checked for impedance, polarity, and voltage ratio before installation.

TRANSFORMER CONNECTIONS

Series Connections of Low-Voltage Windings *(continued)*

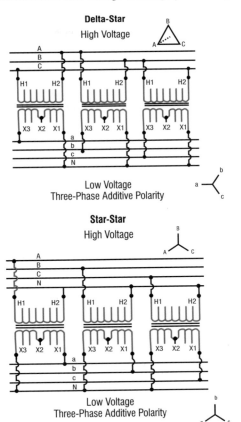

Delta-Star

High Voltage

Low Voltage
Three-Phase Additive Polarity

Star-Star

High Voltage

Low Voltage
Three-Phase Additive Polarity

Note: For additive polarity, the H1 and the X1 bushings are diagonally opposite to each other.

67

Series Connections of Low-Voltage Windings *(continued)*

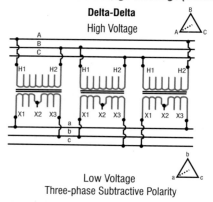

Delta-Delta

High Voltage

Low Voltage
Three-phase Subtractive Polarity

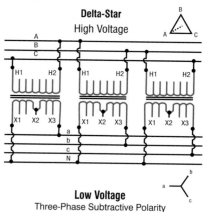

Delta-Star

High Voltage

Low Voltage
Three-Phase Subtractive Polarity

Note: For subtractive polarity the H1 and the X1 bushings are directly opposite each other.

Two Three-Way Switches

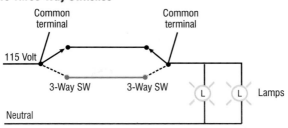

Two Three-Way Switches and One Four-Way Switch

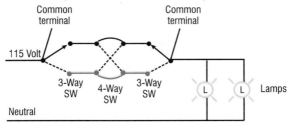

Bell Circuit

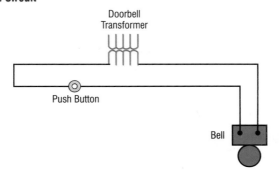

MISCELLANEOUS WIRING DIAGRAMS

Remote-Control Circuit One Relay and One Switch

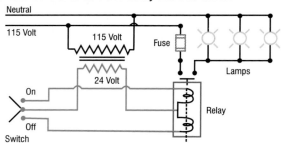

 SUPPORTS FOR RI6ID METAL CONDUIT

Conduit Size	Distance Between Supports
½"–¾"	10 Feet
1"	12 Feet
1¼"–1½"	14 Feet
2"-2½	16 Feet
3" and larger	20 Feet

Source: NFPA 70®, *National Electrical Code*®, 2023 edition, NFPA Quincy, MA, 2022, Table 344.30(B), as modified.

 SUPPORTS FOR RIGID NONMETALUC CONDUIT

Conduit Size	Distance Between Supports
½"–1"	3 Feet
1¼"–2"	5 Feet
2½"–3"	6 Feet
3½"–5"	7 Feet
6	8 Feet

For SI units: (Supports) 1 foot = 0.3048 meter
Source: NFPA 70®, *National Electrical Code*®, 2023 edition, NFPA Quincy, MA, 2022, Table 352.30(B), as modified.

CONDUCTOR PROPERTIES

Size AWG/ Kcmil	Area Cir. Mills.	Conductors				DC Resistance at 75°c (167°F)		
		Standing		Cueral		Copper		Aluminium
		Quan-tity	Diam. (in.)	Diam. (in.)	Area (in.²)	Uncoated onm/k FT	Coated onm/k FT	ohm/ kFT
18	1620	1	---	0.040	0.001	7.77	8.08	12.8
18	1620	7	0.015	0.045	0.002	7.95	6.45	13.1
16	2580	1	---	0.051	0.002	4.89	5.08	8.05
16	2580	7	0.019	0.058	0.003	4.99	5.29	8.21
14	4110	1	---	0.064	0.003	3.07	3.19	5.06
14	4110	7	0.024	0.073	0.004	3.14	3.26	5.17
12	6530	1	---	0.081	0.005	1.93	2.01	3.18
12	6530	7	0.030	0.092	0.006	1.98	2.05	3.25
10	10380	1	---	0.102	0.008	1.21	1.26	2.00
10	10380	7	0.038	0.116	0.011	1.24	1.29	2.04
8	16510	1	---	0.128	0.013	0.764	0.786	1.26
8	16510	7	0.049	0.146	0.017	0.778	0.809	1.28
6	26240	7	0.061	0.184	0.027	0.491	0.510	0.808
4	41740	7	0.077	0.232	0.042	0.308	0.321	0.508
3	52620	7	0.087	0.260	0.053	0.245	0.254	0.403
2	83600	7	0.097	0.292	0.067	0.194	0.201	0.319
1	83690	19	0.066	0.332	0.087	0.154	0.160	0.253
1/0	105600	19	0.074	0.372	0.109	0.122	0.127	0.201
2/0	133100	19	0.084	0.418	0.137	0.0967	0.101	0.159
3/0	167800	19	0.094	0.470	0.173	0.0766	0.0797	0.126
4/0	211600	19	0.106	0.528	0.219	0.0608	0.0626	0.100
250	---	37	0.082	0.575	0.260	0.0515	0.0535	0.0647
300	---	37	0.090	0.630	0.312	0.0429	0.0446	0.0707
350	---	37	0.097	0.681	0.364	0.0367	0.0332	0.0605
400	---	37	0.104	0.728	0.416	0.0321	0.0331	0.0529
500	---	37	0.116	0.813	0.519	0.0258	0.0265	0.0424
600	---	61	0.099	0.893	0.626	0.0214	0.0223	0.0353
700	---	61	0.107	0.964	0.730	0.0134	0.0189	0.0303
750	---	61	0.111	0.998	0.782	0.0171	0.0176	0.0282
800	---	61	0.114	1.030	0.834	0.0161	0.0166	0.0265
900	---	61	0.122	1.094	0.940	0.0143	0.0147	0.0235
1000	---	61	0.128	1.152	1.042	0.0129	0.0132	0.0212
1250	---	91	0.117	1.289	1.305	0.0103	0.0106	0.0169
1500	---	91	0.128	1.412	1.566	0.00858	0.00883	0.0141
1750	---	127	0.117	1.526	1.829	0.00753	0.00756	0.0121
2000	---	127	0.126	1.632	2.052	0.00643	0.00662	0.0106

These resistance values are valid ONLY for the parameters as given. Using conductors having coated stands, different standing type, and, especially, other temperatures changes the resistance.

Equation for temperature change: $R_2 = R_1 [1 + \alpha(T_2 - 75)]$ where $\alpha_{cu} = 0.00323$, $\alpha_{AL} = 0.00330$ at 75°C (167°F).

See *NEC* Chapter 9, Table 8. See *Ugly's* pages 139-149 for metric conversions.

Source: NFPA 70®, *National Electrical Code*®, 2023 edition, NFPA, Quincy, MA, 2022, Chapter 9, Table 8, as modified.

AC RESISTANCE AND REACTANCE FOR 600-VOLT CABLES. THREE-PHASE, 60-HZ. 75°C (167°F): THREE SINGLE CONDUCTORS IN CONDUIT

Size AWG/Kcmil	X_L (Reactance) for All Wires		AC Resistance for Uncoated Copper Wires			AC Resistance for Aluminum Wires			Effective Z at 0.85 PF for Uncoated Copper Wires			Effective Z at 0.85 PF for Aluminium Wires			Size AWG/Kcmil
	PVC Al Conduit	Steel Conduit	PVC Conduit	Al Conduit	Steel Conduit	PVC Conduit	Al Conduit	Steel Conduit	PVC Conduit	Al Conduit	Steel Conduit	PVC Conduit	Al Conduit	Steel Conduit	
14	0058	0073	3.1	3.1	3.1	-	-	-	2.7	2.7	2.7	-	-	-	14
12	0054	0068	2.0	2.0	2.0	3.2	3.2	3.2	1.7	1.7	1.7	2.8	28	2.8	12
10	0050	0063	1.2	1.2	1.2	2.0	2.0	2.0	1.1	1.1	1.1	1.8	1.8	1.8	10
8	0052	0065	0.78	0.78	0.78	1.3	1.3	1.3	0.69	0.69	0.70	1.1	1.1	1.1	8
6	0051	0064	0.49	0.49	0.49	0.81	0.81	081	0.44	0.45	0.45	0.71	0.72	0.72	6
4	0048	0060	0.31	0.31	0.31	0.51	0.51	0.51	0.29	0.29	0.30	0.46	0.46	0.46	4
3	0047	0059	0.25	0.25	0.25	0.40	0.41	0.40	0.23	0.24	0.24	0.37	0.37	0.37	3
2	0045	0057	0.19	0.20	0.20	0.32	0.32	0.32	0.19	0.19	0.20	0.30	0.30	0.30	2
1	0046	0057	0.15	0.16	0.16	0.25	0.26	0.25	0.16	0.16	0.16	0.24	0.24	0.25	1
1/0	0044	0055	0.12	0.13	0.12	0.20	0.21	0.20	0.13	0.13	0.13	0.19	0.20	0.20	1/0
2/0	0043	0054	0.10	0.10	0.10	0.16	0.16	0.16	0.11	0.11	0.11	0.16	0.16	0.16	2/0
3/0	0042	0052	0.077	0.082	0.079	0.13	0.13	0.13	0.088	0.092	0.094	0.13	0.13	0.14	3/0
4/0	0041	0051	0.062	0.067	0.063	0.10	0.11	0.10	0.074	0.078	0.080	0.11	0.11	0.11	4/0
250	0041	0052	0.052	0.057	0.054	0.085	0.090	0.086	0.066	0.070	0.073	0.094	0.098	0.10	250
300	0041	0051	0.044	0.049	0.045	0.071	0.076	0.072	0.059	0.063	0.065	0.082	0.086	0.088	300
350	0040	0050	0.038	0.043	0.039	0.061	0.066	0.063	0.053	0.058	0.060	0.073	0.077	0.080	350
400	0040	0049	0.033	0.038	0.035	0.054	0.059	0.055	0.049	0.053	0.056	0.066	0.071	0.073	400
500	0039	0048	0.027	0.032	0.029	0.043	0.048	0.045	0.043	0.048	0.050	0.057	0.061	0.064	500
600	0039	0048	0.023	0.028	0.025	0.016	0041	0.038	0040	0.044	0.047	0.051	0.055	0.058	600
750	0038	0048	0.019	0.024	0.021	0.029	0.034	0.031	0.036	0.040	0.043	0.045	0.049	0.052	750
1000	0037	0046	0.015	0.019	0.018	0.023	0.027	0.025	0.032	0036	0.040	0.039	0.042	0.046	1000

Notes: See NEC Chapter 9, Table 9 for assumptions and explanations. See *Ugly's* pages 139–140 for metric conversions.
Source: NFPA 70®, *National Electrical Code*®, 2023 edition, NFPA, Quincy, MA, 2022, Chapter 9, Table 9, as modified.

AMPACITIES OF INSULATED CONDUCTORS

Ampacities of Insulated Conductors with Not More Than Three Current-Carrying Conductors in Raceway, Cable, or Earth (Directly Buried)

Size	60°C (140°F)	75°C (167°F)	90°C (194°F)	60°C (140°F)	75°C (167°F)	90°C (194°F)	Size
AWG or kcmil	Types TW UF	Types RHW, THHW, THW, THWN, XHHW, XHWN, USE ZW	Types TBS, SA, SIS, FEP FEPB, MI, PFA, RHH.RHW-2,THHN THHW, THW-2, THWN-2,USE-2, XHH, XHHW, XHHW-2, XHWN, XHWN-2, XHHW-2 ZW-2	Types TW UF	Types RHW, THHW, THW, THWN, XHHW, XHWN, USE	Types TBS..SA. SIS, THHN, THHW, THW-2,THWN-2, RHH, RHW-2, USE-2, XHH, XHHW, XHHW -2, XHWN-2, XHHN	AWG or kcmil
	Copper			Aluminum or Copper Clad Aluminum			
14*	15	20	25	------	------	------	
12*	20	25	30	15	20	25	12*
10*	30	35	40	25	30	35	10*
8	40	50	55	35	40	45	8
6	55	65	75	40	50	55	6
4	70	85	95	55	65	75	4
3	85	100	115	65	75	85	3
2	95	115	130	75	90	100	2
1	110	130	145	85	100	115	1
1/0	125	150	170	100	120	135	1/0
2/0	145	175	195	115	135	150	2/0
3/0	165	200	225	130	155	175	3/0
4/0	195	230	260	150	180	205	4/0
250	215	255	290	170	205	230	250
300	240	285	320	195	230	260	300
350	260	310	350	210	250	280	350
400	280	335	380	225	270	305	400
500	320	380	430	260	310	350	500
600	350	420	475	285	340	385	600
700	385	460	520	315	375	425	700
750	400	475	535	320	385	435	750
800	410	490	555	330	395	445	800
900	435	520	585	355	425	480	900
1000	455	545	615	375	445	500	1000
1250	495	590	665	405	485	545	1250
1500	525	625	705	435	520	585	1500
1750	545	650	735	455	545	615	1750
2000	555	665	750	470	560	630	2000

Notes:
1. Section 310.15(B) shall be referenced for ampacity correction factors where the ambient temperature is other than 30°C (86°F).
2. Section 310.15(C)(1) shall be referenced for more than three current-carrying conductors.
3. Section 310.16 shall be referenced for conditions of use.
*Section 240.4(D) shall be referenced for conductor overcurrent protection limitations, except as modified elsewhere in the *Code*.
See *Ugly's* page 77 for Adjustment Examples®.
Source: NFPA 70®, *National Electrical Code*®, 2023 edition, NFPA, Quincy, MA, 2022, Table 310.16, as modified.

AMPACITIES OF INSULATED CONDUCTORS

Ampacities of Single-Insulated Conductors in Free Air

Size	60°C (140°F)	75°C (167°F)	90°C (194°F)	60°C (140°F)	75°C (167°F)	90°C (194°F)	Size
AWG or kcmil	Types TW UF	Types RHW THHW. THW, THWN, XHHW, XHWN, ZW	Types TBS...SA, SIS, FEP FEPB, MI, PFA, RHH, RHW-2, THHN, THHW, THW-2, THWN-2, USE-2 XHH, XHHW XHHW-2, XHWN, XHWN-2, XHWN, 2, ZW-2	Types TW UF	Types RHW THHW. THW, THWN, XHHW, XHWN	Types TBS...SA. SIS, THHN, THHW, THW-2.THWN-2, RHH, RHW-2, USE-2, XHH, XHHW, XHHW -2, XHWN, XHWN-2, XHHN	AWG or kcmil
			Copper		Aluminum or Copper Clad Aluminum		
14*	25	30	35	----	----	----	----
12*	30	35	40	25	30	35	12*
10*	40	50	55	35	40	45	10*
8	60	70	80	45	55	60	8
6	80	95	105	60	75	85	6
4	105	125	140	80	100	115	4
3	120	145	165	95	115	130	3
2	140	170	190	110	135	150	2
1	165	195	220	130	155	175	1
1/0	195	230	260	150	180	205	1/0
2/0	225	265	300	175	210	235	2/0
3/0	260	310	350	200	240	270	3/0
4/0	300	360	405	235	280	315	4/0
250	340	405	455	265	315	355	250
300	375	445	500	290	350	395	300
350	420	505	570	330	395	445	350
400	455	545	615	355	425	480	400
500	515	620	700	405	485	545	500
600	575	690	780	455	540	615	600
700	630	755	850	500	595	670	700
750	655	785	885	515	620	700	750
800	680	815	920	535	645	725	800
900	730	870	980	580	700	790	900
1000	780	935	1055	625	750	845	1000
1250	890	1065	1200	710	855	965	1250
1500	980	1175	1325	795	950	1070	1500
1750	1070	1280	1445	875	1050	1185	1750
2000	1155	1385	1560	960	1150	1295	2000

Notes:
1. Section 310.15(B) shall be referenced for ampacity correction factors where the ambient temperature is other than 30°C (86°F).
2. Section 310.17 shall be referenced for conditions of use.
*Section 240.4(D) shall be referenced for conductor overcurrent protection limitations, except as modified elsewhere in the *Code.*
See *Ugly's* page 77 for Adjustment Examples.
Source: NFPA 70®, *National Electrical Code*®, 2023 edition, NFPA, Quincy, MA, 2022. Table 310.17, as modified.

AMPACITIES OF INSULATED CONDUCTORS

Ampacities of Insulated Conductors with Not More Than Three
Current-Carrying Conductors in Raceway or Cable

Size	150°C (302°F)	200°C (392°F)	250°C (482°F)	150°C (302°F)	Size
AWG or kcmil	Type Z	Types FEP FEPS PFA SA	Types PFAH TFE	Type Z	AWG or kcmil
	Copper		Nickel or Nickel-Coated Copper	Aluminium or Copper-Clad Aluminium	
14	34	36	39	----	14
12	43	45	54	30	12
10	55	60	73	44	10
8	76	83	93	57	8
6	96	110	117	75	6
4	120	125	148	94	4
3	143	152	166	109	3
2	160	171	191	124	2
1	186	197	215	145	1
1/0	215	229	244	169	1/0
2/0	251	260	273	198	2/0
3/0	288	297	308	227	3/0
4/0	332	346	361	260	4/0

Notes:
1. Section 310.15(B) shall be referenced for ampacity correction factors where the ambient temperature is other than 40°C (104°F).
2. Section 310.15(C)(1) shall be referenced for more than three current-carrying conductors.
3. Section 310.18 shall be referenced for conditions of use.
Source: NFPA 70®, *National Electrical Code®*, 2023 edition, NFPA. Quincy. MA, 2022. Table 310.18 as modified.

AMPACITIES OF INSULATED CONDUCTORS

Ampacities of Single-Insulated Conductors in Free Air

Size	150°C (302°F)	200°C (392°F)	250°C (482°F)	150°C (302°F)	Size
AWG or kcmil	Type Z	Types FEP FEPS PFA SA	Types PFAH TFE	Type Z	AWG or kcmil
	Copper		Nickel or Nickel-Coated Copper	Aluminium or Copper-Clad Aluminium	
14	46	54	59	---	14
12	60	68	78	47	12
10	80	90	107	63	10
8	106	124	142	83	8
6	155	165	205	112	6
4	190	220	278	148	4
3	214	252	327	170	3
2	255	293	381	198	2
1	293	344	440	228	1
1/0	339	399	532	263	1/0
2/0	390	467	591	305	2/0
3/0	451	546	708	351	3/0
4/0	529	629	830	411	4/0

Notes:
1. Section 310.15(B) shall be referenced for ampacity correction factors where the ambient temperature is other than 40°C (104°F).
2. Section 310.19 shall be referenced for conditions of use.
Source: NFPA 70®, *National Electrical Code*®, 2023 edition, NFPA. Quincy. MA, 2022. Table 310.19 as modified.

AMPACITY CORRECTION AND ADJUSTMENT FACTORS

Examples

Ugly's page 73 shows ampacity values for not more than three current-carrying conductors in a raceway or cable and the wiring installed in a 30°C (86 °F) ambient temperature.

Example 1: A raceway contains three 3 AWG THWN conductors for a three-phase circuit at an ambient temperature of <u>30°C (86°F).</u> *Ugly's* page 73, 75° C (167°F) column indicates **100 amps.**

Example 2: A raceway contains three 3 AWG THWN conductors for a three-phase circuit at an ambient temperature of <u>40°C (104°F).</u> *Ugly's* page 73, 75°C (167°F) column indicates **100 amps.** This value must be corrected for ambient temperature (see note on Temperature Correction Factors at bottom of *Ugly's* page 73). 40°C (104°F) factor is **0.88.**
100 Amps x 0.88 = **88 Amps** =corrected ampacity

Example 3: A raceway contains six 3 AWG THWN conductors for two three-phase circuits at an ambient temperature of 30°C (86°F).
Ugly's page 73, 75°C (167°F) column indicates **100 amps.** This value must be adjusted for more than three current-carrying conductors. The table on *Ugly's* page 78 requires an adjustment of 80% for four through six current-carrying conductors.
100 Amps x 80% = **80 amps**
The adjusted ampacity is **80 Amps.**

Example 4: A raceway contains six 3 AWG THWN conductors for two three-phase circuits in an ambient temperature of 40°C (104°F). These conductors must be corrected and adjusted (or derated) for both ambient temperature and number of current-carrying conductors.
Ugly's page 73, 75°C (167°F) column indicates **100 amps.**
NEC Table 310.15(B)(1)(1), 40°C(104°F) temperature factor is **0.88.**
Ugly's page 78,4 - 6 conductor factor is **0.80.**
100 Amps x 0.88 x 0.80 = **70.4 Amps**
The new derated ampacity is **70.4 Amps.**

 **ADJUSTMENT FACTORS**

For More Than Three Current-Carrying Conductors in a Raceway or Cable

Number of Current-Carrying Conductors*	Percent of Values in Tables 310.16 Through 310. 19 as Adjusted for Ambient Temperature if Necessary
4-6	80
7-9	70
10-20	50
21-30	45
31-40	40
41 and above	35

*Number of conductors is the total number of conductors in the raceway or cable, including spare conductors. The count shall be adjusted in accordance with 310.15(E) and (F). The count shall not include conductors that are connected to electrical components that cannot be simultaneously energized.

Source: NFPA 70®, *National Electrical Code®*, 2023 edition, NFPA, Quincy, MA, 2022, Table 310.15(C)(1).

Conductor and Equipment Termination Ratings*

Examples:

A 150-amps circuit breaker is labeled for 75°C (167°F) terminations and is selected to be used for a 145-amps, noncontinuous load. It would be permissible to use a 1/0 THWN conductor that has a 75°C (167°F) insulation rating and has an ampacity of 150 amps (*Ugly's* page 73).

When a THHN (90°C [194°F]) conductor is connected to a 75°C (167°F) termination, it is limited to the 75°C (167°F) ampacity. Therefore, if a 1 THHN conductor with a rating of 145 amps were connected to a 75°C (167°F) terminal, its ampacity would be limited to 130 amps instead of 150 amps, which is too small for the load (*Ugly's* page 73).

If the 145 amp, noncontinuous load listed above uses 1/0 THWN conductors rated 150 amps and the conductors are in an ambient temperature of 40°C (104°F), the conductors would have to be corrected for the ambient temperature.

⚡ ADJUSTMENT FACTORS

From *NEC* Table 310.15(B)(1)(1), 40°C (104°F) ambient temperature correction factor = **0.88**
1/0 THWN = 150 Amps
150 Amps x 0.88 = **132 Amps** (which is too small for the 145-amp load, so a larger size conductor is required).

Apply temperature correction factors to the next size THWN conductor.
2/0 THWN = 175 Amps (from the 75°C [167°F] column—*Ugly's* page 73)
175 Amps x 0.88 = **154 Amps**. This size is suitable for the 145-amps load.

The advantage of using 90°C (194°F) conductors is that you can apply ampacity derating factors to the higher 90°C (194°F) ampacity rating, and it may save you from going to a larger conductor.

1/0 THHN = 170 Amps (from the 90°C [194°F] column—*Ugly's* page 73).
40°C [104°F] ambient temperature correction factor for a 90°C conductor = 0.91 (*NEC* Table 310.15(B)(1)(1))
170 Amps x 0.91 = **154.7 Amps**.
This size is suitable for the 145-amp load.
This 90°C [194°F] conductor can be used but can never have a final derated ampacity over the rating of 1/0 THWN 75°C [167°F] rating of 150 amps.

You are allowed to use higher temperature (insulated conductors) such as THHN (90°C [194°F]) conductors on 60°C (140°F) or 75°C (167°F) terminals of circuit breakers and equipment, and you are allowed to derate from the higher value for temperature and number of conductors, but the final derated ampacity is limited to the 60°C or 75°C (140°F or 167°F) terminal insulation labels.

*See *NEC* 2023 Section 110.14(C)(1) and (2).

CONDUCTOR APPLICATIONS AND INSULATIONS

Trade Name	Letter	Max. Temp.	Application Provisions
Fluorinated Ethylene Propylene	FEP or FEPB	90°C (194°F)	Dry and Damp Locations
		200°C (392°F)	Dry Locations—Special Applications[1]
Mineral Insulation (Metal Sheathed)	MI	90°C (194°F)	Dry and Wet Locations
		250°C (482°F)	Special Applications[1]
Moisture-, Heat-, and Oil-Resistant Thermoplastic	MTW	60°C (140°F)	Machine Tool Wiring in Wet Locations
		90°C (194°F)	Machine Tool Wiring in Dry Locations, Informational Note: See NFPA 79
Paper		85°C (185°F)	For Underground Service Conductors, or By Special Permission
Perfluoro-Alkoxy	PFA	90°C (194°F)	Dry and Damp Locations
		200°C (392°F)	Dry Locations—Special Applications[1]
Perfluoro-Alkoxy	PFAH	250°C (482°F)	Dry Locations Only. Only For Leads Within Apparatus or Within Raceways Connected to Apparatus (Nickel or Nickel-Coated Copper Only)
Thermoset	RHH	90°C (194°F)	Dry and Damp Locations
Moisture-Resistant Thermoset	RHW	75°C (167°F)	Dry and Wet Locations
Moisture-Resistant Thermoset	RHW-2	90°C (194°F)	Dry and Wet Locations
Silicone	SA	90°C (194°F)	Dry and Damp Locations
		200°C (392°F)	Special Applications[1]
Thermoset	SIS	90°C (194°F)	Switchboard and Switchgear Wiring Only
Thermoplastic and Fibrous Outer Braid	TBS	90°C (194°F)	Switchboard and Switchgear Wiring Only

See *Ugly's* page 82 for footnotes on special provisions and/or applications.

(continued on next page)

Trade Name	Letter	Max. Temp.	Application Provisions
Extended Polytetrafluoro-Ethylene	TFE	250°C (482°F)	Dry Locations Only, Only tor Leads Within Apparatus or Within Raceways Connected to Apparatus, or as Open Wiring (Nickel or Nickel-Coated Copper Only
Heat-Resistant Thermoplastic	THHN	90°C (194°F)	Dry and Damp Locations
Moisture- and Heat-Resistant Thermoplastic	THHW	75°C (167°F)	Wet Location
		90°C (194°F)	Dry Location
Moisture-and Heat-Resistant Thermoplastic	THW	75°C (167°F)	Dry and Wet Locations
		90°C (194°F)	Special Appl. Within Electric Discharge Lighting Equipment, Limited to 1000 Open-Circuit Volts or Less, (Size 14-8 Only as Permitted in Section 410,68
	THW-2	90°C (194°F)	Dry and Wet Locations
Moisture-and Heat-Resistant Thermoplastic	THWN	75°C (167°F)	Dry and Wet Locations
	THWN-2	90°C (194°F)	
Moisture-Resistant Thermoplastic	TW	60°C (140°F)	Dry and Wet Locations
Underground Feeder and Branch-Circuit Cable Single Conductor (For Type "UF" Cable Employing More Than 1 Conductor, (See Part II of Article 340)	UF	60°C (140°F) 75°C (167°F)[2]	See *NEC* Article 340, Part II

See *Ugly's* page 82 for footnotes on special provisions and/or applications.

(continued on next page)

CONDUCTOR APPLICATIONS AND INSULATIONS

Trade Name	Letter	Max. Temp.	Application Provisions
Underground Service-Entrance Cable Single Conductor (For Type "USE" Cable Employing More Than 1 Conductor. See Part II of Article 338)	USE	75°C (167°F)[2]	See *NEC* Article 338, Part II
	USE-2	90°C (194°F)	Dry and Wet Locations
Thermoset	XHH	90°C (194°F)	Dry and Damp Locations
Flame-Retardant Thermoset	XHHN	90°C (194°F)	Dry and Damp Locations
Moisture-Resistant Thermoset	XHHW	90°C (194°F)	Dry and Damp Locations
		75°C (167°F)	Wet Locations
Moisture-Resistant, Thermoset	XHHW-2	90°C (194°F)	Dry and Wet Locations
Flame-Retardant, Moisture-Resistant Thermoset	XHWN	75°C (167°F)	Dry and Wet Locations
	XHWN-2	90°C (194°F)	
Modified Ethylene Tetrafluoro-Ethylene	Z	90°C (194°F)	Dry and Damp Locations
		150°C (302°F)	Dry Locations—Special Applications[1]
Modified Ethylene Tetrafluoro-Ethylene	ZW	75°C (167°F)	Wet Locations
		90°C (194°F)	Dry and Damp Locations
		150°C (302°F)	Dry Locations—Special Applications[1]
	ZW-2	90°C (194°F)	Dry and Wet Locations

1 For signaling circuits permitting 300-volt insulation.
2 For ampacity limitation, see 340.80 *NEC*.

Source: NFPA 70®, *National Electrical Code*®, 2023 edition, NFPA, Quincy, MA, 2022, Table 310.4(1), as modified.

Note: Some insulations do not require an outer covering.

MAXIMUM NUMBER OF CONDUCTORS IN TRADE SIZES OF CONDUIT OR TUBING

The *National Electrical Code ®* shows a separate table for each type of conduit. In order to keep *Ugly's Electrical References* in a compact and easy-to-use format, the following tables are included:

Electrical Metallic Tubing (EMT). Electrical NonmetaKc Tubing (ENT), PVC 40, PVC 80, Rigid Metal Conduit. Flexible Metal Conduit, and Liquidtight Flexible Metal Conduit

When other types of conduit are used, refer to *NEC* Inftxmatwe Annex C or use method shown below to figure conduit size.

Example #1: All same wire size and type Insulation.

10 – 12 AWG RHH (With outer covering) in Intermediate Metal Conduit.
Go to the RHH Conductor Square Inch Area Table. (Ugly's page 98)
12 AWG RHH = 0.0353 sq. in.   10 x 0.0353 sq. in. = 0.353 sq. in.
Go to Intermediate Metal Conduit Square Inch Area Table. (Ugly's page 101)
Use "Over 2 Wires 40%' column.
¾ -inch conduit = 0.235 sq. in, (less than 0,353, so its too small),
1-inch conduit = 0.384 sq. in. (greater than 0.353, so it's correct size).

Example #2 Different wire sizes or types Insulation.

10 – 12 AWG RHH (with outer covering and 10 - 10 AWG THHN in Liquidtight Nonmetallic Conduit (LFNC-B).
Go to the RHH Conductor Square Inch Area Table. (Ugly's page 98)
12 AWG RHH = 0,0353 sq. in.   10 x 0.0353 sq. in. = 0.353 sq. in.
Go to the THHN Conductor Square Inch Area Table. (Ugly's page 98)
10 AWG THHN = 0.0211 sq, in,   10 x 0.0211 sq. in. = 0.211 sq. in.
0.353 sq, in. + 0.211 sq. in. = 0,564 sq, in.
Go to Uquidtight Flexible Nonmetallic Conduit (LFNC-B) Square Inch Table. (Ugly's *page 102*)
Use "Over 2 Wires 40%" column.
1-inch conduit = 0.349 sq. in. (less than 0.564, so it's too small)
1¼ inch conduit = 0.611 sq. in. (greater than 0.564, so it's the correct size).

Note 1*: All conductors must be counted including grounding conductors for fillpercentage.
Note 2: When all conductors are of same type and size, decimals 0.8 and larger can be rounded up.
Note 3** : These are minimum size calculations, under certain conditions jamming can occur and the next size conduit must be used.
Note 4***: **Caution** —When over three current carrying conductors are used in same circuit, conductor ampacity must be lower (adjusted).

*   See Appendix C and Chapter 9 2023 *NEC*. for complete tables and examples.
**   See Chapter 9 Table 1 and Notes to Tables 1- 9. 2023 *NEC*.
*** See 2020 *NEC* 310.15 for adjustment factors for temperature and number of current-carrying conductors.

MAXIMUM NUMBER OF CONDUCTORS IN ELECTRICAL METALLIC TUBING

Type Letters	Cond.Size AWG/kcmil	½	¾	1	1¼	1½	2	2½	3	3½	4	5	6
RHH, RHW, RHW-2	14	4	7	11	20	27	46	80	120	157	201	302	427
	12	3	6	9	17	23	38	66	100	131	167	251	354
	10	2	5	8	13	18	30	53	81	105	135	203	286
	8	1	2	4	7	9	16	28	42	55	70	106	150
	6	1	1	3	5	8	13	22	34	44	56	85	120
	4	1	1	2	4	6	10	17	26	34	44	66	94
	3	1	1	1	3	5	9	15	23	30	38	58	82
	2	1	1	1	3	4	7	13	20	26	33	50	71
	1	0	1	1	1	3	5	9	13	17	22	33	47
	1/0	0	1	1	1	2	4	7	11	15	19	29	41
	2/0	0	1	1	1	2	4	6	10	13	17	25	35
	3/0	0	0	1	1	1	3	5	8	11	14	21	30
	4/0	0	0	1	1	1	3	5	7	9	12	18	26
	250	0	0	0	1	1	1	3	6	7	9	14	20
	300	0	0	0	1	1	1	3	5	6	8	12	17
	350	0	0	0	1	1	1	3	4	6	7	11	16
	400	0	0	0	1	1	1	2	4	5	7	10	14
	500	0	0	0	0	1	1	1	3	4	6	8	12
	600	0	0	0	0	1	1	1	2	3	4	7	10
	700	0	0	0	0	0	1	1	1	3	4	6	9
	750	0	0	0	0	0	1	1	1	2	3	4	8
TW, THHW, THW, THW-2	14	8	15	25	43	58	96	168	254	332	424	638	900
	12	6	11	19	33	45	74	129	195	255	326	490	691
	10	5	8	14	24	33	55	96	145	190	243	365	515
	8	2	5	8	13	18	30	53	81	105	135	203	286
RHH*, RHW*, RHW-2*	14	6	10	16	28	39	64	112	169	221	282	424	599
	12	4	8	13	23	31	51	90	136	177	227	341	481
	10	3	6	10	18	24	40	70	106	138	177	266	376
	8	1	4	6	10	14	24	42	63	83	106	159	225
RHH*, RHW*, RHW-2*, TW, THW, THHW, THW-2	6	1	3	4	8	11	18	32	48	63	81	122	172
	4	1	1	3	6	8	13	24	36	47	60	91	128
	3	1	1	3	5	7	12	20	31	40	52	78	110
	2	1	1	2	4	6	10	17	26	34	44	66	94
	1	1	1	1	3	4	7	12	18	24	31	46	66
	1/0	0	1	1	2	3	6	10	16	20	26	40	56
	2/0	0	1	1	1	3	5	9	13	17	22	33	47
	3/0	0	1	1	1	2	4	7	11	15	19	28	40
	4/0	0	0	1	1	1	3	6	9	12	16	24	33
	250	0	0	1	1	1	3	5	7	10	13	19	27
	300	0	0	1	1	1	2	4	6	8	11	16	23
	350	0	0	0	1	1	1	4	6	7	9	14	21
	400	0	0	0	1	1	1	3	5	7	9	13	19
	500	0	0	0	1	1	1	3	4	6	7	11	16
	600	0	0	0	1	1	1	2	3	4	6	9	13
	700	0	0	0	0	1	1	1	3	4	5	8	11
	750	0	0	0	0	1	1	1	3	4	5	7	10
THHN, THWN, THWN-2	14	12	22	35	61	84	138	241	364	476	608	914	1290
	12	9	16	26	45	61	101	176	266	347	443	666	941
	10	5	10	16	28	38	63	111	167	219	279	420	593
	8	3	6	9	16	22	36	64	96	126	161	242	342
	6	2	4	7	12	16	26	46	69	91	116	175	247
	4	1	2	4	7	10	16	28	43	56	71	107	152
	3	1	1	3	6	8	13	24	36	47	60	91	128
	2	1	1	3	5	7	11	20	30	40	51	76	108
	1	1	1	1	4	5	8	15	22	29	37	56	80

(continued on next page)

84

MAXIMUM NUMBER OF CONDUCTORS IN ELECTRICAL METALLIC TUBING

Type Letters	Cond. Size AWG/kcmil	Trade Sizes in Inches											
		½	¾	1	1¼	1½	2	2½	3	3½	4	5	6
THHN, THWN, THWN-2	1/0	1	1	1	3	4	7	12	19	25	32	47	67
	2/0	0	1	1	2	3	6	10	16	20	26	40	56
	3/0	0	1	1	1	3	5	8	13	17	22	33	46
	4/0	0	1	1	1	2	4	7	11	14	18	27	38
	250	0	0	1	1	1	3	6	9	11	15	22	31
	300	0	0	1	1	1	3	5	7	10	13	19	27
	350	0	0	1	1	1	2	4	6	9	11	17	24
	400	0	0	0	1	1	1	4	6	8	10	15	21
	500	0	0	0	1	1	1	3	5	6	8	12	17
	600	0	0	0	1	1	1	2	4	5	7	10	14
	700	0	0	0	1	1	1	2	3	4	6	9	12
	750	0	0	0	0	1	1	1	3	4	5	8	12
FEP, FEPB, PFA, PFAH, TFE	14	12	21	34	60	81	134	234	354	462	590	886	1252
	12	9	15	25	43	59	98	171	258	337	430	647	913
	10	6	11	18	31	42	70	122	185	241	309	464	655
	8	3	6	10	18	24	40	70	106	138	177	266	376
	6	2	4	7	12	17	28	50	75	98	126	189	267
	4	1	3	5	9	12	20	35	53	69	88	132	187
	3	1	2	4	7	10	16	29	44	57	73	110	155
	2	1	1	3	6	8	13	24	36	47	60	91	128
PFA, PFAH, TFE	1	1	1	2	4	6	9	16	25	33	42	63	89
PFA, PFAH, TFE, Z	1/0	1	1	1	3	5	8	14	21	27	35	53	74
	2/0	0	1	1	3	4	6	11	17	22	29	43	61
	3/0	0	1	1	2	3	5	9	14	18	24	36	51
	4/0	0	1	1	1	2	4	8	11	15	19	29	41
Z	14	14	25	41	72	98	161	282	426	556	711	1068	1508
	12	10	18	29	51	69	114	200	302	394	504	758	1070
	10	6	11	18	31	42	70	122	185	241	309	464	655
	8	4	7	11	20	27	44	77	117	153	195	293	414
	6	3	5	8	14	19	31	54	82	107	137	206	291
	4	1	3	5	9	13	21	37	56	74	94	142	200
	3	1	2	4	7	9	15	27	41	54	69	103	146
	2	1	1	3	6	8	13	22	34	45	57	86	121
	1	1	1	2	4	6	10	18	28	36	46	70	98
XHH, XHHW, XHHW-2, ZW	14	8	15	25	43	58	96	168	254	332	424	638	900
	12	6	11	19	33	45	74	129	195	255	326	490	691
	10	5	8	14	24	33	55	96	145	190	243	365	515
	8	2	5	8	13	18	30	53	81	105	135	203	286
	6	1	3	6	10	14	22	39	60	78	100	150	212
	4	1	2	4	7	10	16	28	43	56	72	109	153
	3	1	1	3	6	8	14	24	36	48	61	92	130
	2	1	1	3	5	7	11	20	31	40	51	77	109
XHH, XHHW, XHHW-2	1	1	1	1	4	5	8	15	23	30	38	57	81
	1/0	1	1	1	3	4	7	13	19	25	32	48	68
	2/0	0	1	1	2	3	6	10	16	21	27	40	57
	3/0	0	1	1	1	3	5	9	13	17	22	33	47
	4/0	0	1	1	1	2	4	7	11	14	18	27	39
	250	0	0	1	1	1	3	6	9	12	15	22	32
	300	0	0	1	1	1	3	5	8	10	13	19	27
	350	0	0	1	1	1	2	4	7	9	11	17	24
	400	0	0	0	1	1	1	4	6	8	10	15	21
	500	0	0	0	1	1	1	3	5	6	8	12	18
	600	0	0	0	1	1	1	2	4	5	6	10	14
	700	0	0	0	1	1	1	2	3	4	6	9	12
	750	0	0	0	0	1	1	1	3	4	6	9	12

*Types RHH, RHW, and RHW-2 without outer covering.

See *Ugly's* page 141 for Trade Size/Metric Designator conversion.

Source: NFPA 70®, *National Electrical Code®*, 2023 edition, NFPA, Quincy, MA, 2022, Annex C, Table C.1, as modified.

MAXIMUM NUMBER OF CONDUCTORS IN ELECTRICAL NONMETALLIC TUBING

Type Letters	Cond. Size AWG/kcmil	Trade Sizes in Inches					
		½	¾	1	1¼	1½	2
RHH, RHW RHW-2	14	4	7	11	20	27	45
	12	3	5	9	16	22	37
	10	2	4	7	13	18	30
	8	1	2	4	7	9	15
	6	1	1	3	5	7	12
	4	1	1	2	4	6	10
	3	1	1	1	4	5	8
	2	1	1	1	3	4	7
	1	0	1	1	1	3	5
	1/0	0	1	1	1	2	4
	2/0	0	0	1	1	1	3
	3/0	0	0	1	1	1	3
	4/0	0	0	1	1	1	2
	250	0	0	0	1	1	1
	300	0	0	0	1	1	1
	350	0	0	0	1	1	1
	400	0	0	0	1	1	1
	500	0	0	0	0	1	1
	600	0	0	0	0	1	1
	700	0	0	0	0	0	1
	750	0	0	0	0	0	1
TW, THHW, THW, THW-2	14	8	14	24	42	57	94
	12	6	11	18	32	44	72
	10	4	8	13	24	32	54
	8	2	4	7	13	18	30
RHW,* RHW* RHW-2*	14	5	9	16	28	38	63
	12	4	8	13	22	30	50
	10	3	6	10	17	24	39
	8	1	3	6	10	14	23
RHH,* RHW* RHW-2* TW THW, THHW THW-2	6	1	2	4	8	11	18
	4	1	1	3	6	8	13
	3	1	1	2	5	7	11
	2	1	1	1	4	6	10
	1	0	1	1	3	4	7
	1/0	0	1	1	2	3	8
	2/0	0	1	1	1	3	5
	3/0	0	1	1	1	2	4
	4/0	0	0	1	1	1	3
	250	0	0	1	1	1	3
	300	0	0	1	1	1	2
	350	0	0	0	1	1	1
	400	0	0	0	1	1	1
	500	0	0	0	1	1	1
	600	0	0	0	0	1	1
	700	0	0	0	0	1	1
	750	0	0	0	0	1	1
THHN, THWN THWN-2	14	11	21	34	60	82	135
	12	6	15	25	43	59	99
	10	5	9	15	27	37	62
	8	3	5	9	16	21	36
	6	1	4	6	11	15	26
	4	1	2	4	7	9	16
	3	1	1	3	6	8	13
	2	1	1	3	5	7	11
	1	1	1	1	3	5	8

(continued on next page)

Type Letters	Cond.Size AWG/kcmil	Trade Sizes in Inches					
		½	¾	1	1¼	1½	2
THHN THWW THWW-2	1/0	1	1	1	3	4	7
	2/0	0	1	1	2	3	8
	3/0	0	1	1	1	3	5
	4/0	0	1	1	1	2	4
	250	0	0	1	1	1	3
	300	0	0	0	1	1	3
	350	0	0	1	1	1	2
	too	0	0	0	0	1	1
	500	0	0	0	1	1	1
	600	0	0	0	0	1	1
	700	0	0	0	0	1	1
	750	0	0	0	0	0	1
FEP FEPB PFA PFAH, TFE	14	11	20	33	58	79	131
	12	8	15	24	42	58	96
	10	6	10	17	30	41	69
	8	3	6	10	17	24	39
	6	2	4	7	12	17	28
	4	1	3	5	8	12	19
	3	1	2	4	7	10	16
	2	1	1	3	6	8	13
PFA, PFAH, TFE	1	1	1	2	4	5	9
PFA, PFAH, TFE. Z	1/0	1	1	1	3	4	8
	2/0	0	1	1	3	4	6
	3/0	0	1	1	2	3	5
	4/0	0	1	1	1	2	4
Z	14	13	24	40	70	95	158
	12	9	17	28	49	68	112
	10	6	10	17	30	41	69
	8	3	6	11	19	26	43
	6	2	4	7	13	18	30
	4	1	3	5	9	12	21
	3	1	2	4	6	9	15
	2	1	1	3	5	7	12
	1	1	1	2	4	6	10
XHH, XHHW XHHW-2 ZW	14	8	14	24	42	57	94
	12	6	11	18	32	44	72
	10	4	8	13	24	32	54
	8	2	4	7	13	18	30
	6	1	3	5	10	13	22
	4	1	2	4	7	9	16
	3	1	1	3	6	8	13
	2	1	1	3	3	7	11
XHH XHHW XHHW-2	1	1	1	1	3	5	8
	1/0	0	1	1	3	4	7
	2/0	0	1	1	2	3	6
	3/0	0	1	1	1	3	5
	4/0	0	1	1	1	2	4
	250	0	0	1	1	1	3
	300	0	0	1	1	1	3
	350	0	0	0	1	1	2
	400	0	0	0	1	1	2
	500	0	0	0	0	1	1
	600	0	0	0	0	1	1
	700	0	0	0	0	0	1
	750	0	0	0	0	1	1

*Types RHH, RHW, and RHW-2 without outer covering.

See *Ugly's* page 141 for Trade Size/Metric Designator conversion.

Source: NFPA 70®, *National Electrical Code*®, 2023 edition, NFPA, Quincy, MA, 2022, Annex C, Table C.2, as modified.

MAXIMUM NUMBER OF CONDUCTORS IN RIGID PVC CONDUIT, SCHEDULE 40

Type Letters	Cond.Size AWG/kcmil	Trade Sizes in Inches											
		½	¾	1	1¼	1½	2	2½	3	3½	4	5	6
RHH, RHW. RHW-2	14	4	7	11	20	27	45	64	99	133	171	269	390
	12	3	5	9	16	22	37	53	82	110	142	224	323
	10	2	4	7	13	18	30	43	66	89	115	181	261
	8	1	2	4	7	9	15	22	35	46	60	94	137
	6	1	1	3	5	7	12	18	28	37	48	76	109
	4	1	1	2	4	6	10	14	22	29	37	59	85
	3	1	1	1	4	5	8	12	19	25	33	52	75
	2	1	1	1	3	4	7	10	16	22	28	45	65
	1	0	1	1	1	3	5	7	11	14	19	29	43
	1/0	0	1	1	1	2	4	6	9	13	16	26	37
	2/0	0	0	1	1	1	3	5	8	11	14	22	32
	3/0	0	0	1	1	1	3	4	7	9	12	19	28
	4/0	0	0	0	1	1	2	4	6	8	10	16	24
	250	0	0	0	1	1	1	3	4	6	8	12	18
	300	0	0	0	0	1	1	2	4	5	7	11	16
	350	0	0	0	0	1	1	2	3	5	6	10	14
	400	0	0	0	1	1	1	1	3	4	6	9	13
	500	0	0	0	0	1	1	1	3	4	5	8	11
	600	0	0	0	0	1	1	1	2	3	4	6	9
	700	0	0	0	0	0	1	1	1	3	3	6	8
	750	0	0	0	0	0	1	1	1	3	3	5	8
TW. THHW, THW. THW-2	14	8	14	24	42	57	94	135	209	280	361	568	822
	12	6	11	18	32	44	72	103	160	215	277	436	631
	10	4	8	13	24	32	54	77	119	160	206	325	470
	8	2	4	7	13	18	30	43	66	89	115	181	261
RHH*, RHW,* RHW-2*	14	5	9	16	26	38	63	90	139	186	240	378	546
	12	4	8	13	22	30	50	72	112	150	193	304	439
	10	3	6	10	17	24	39	56	87	117	/0	73/	343
	8	1	3	6	10	14	23	33	52	/0	90	142	205
RHH*. RHW, RHW-2*. TW. THW, THHW. THW-2	6	1	2	4	8	11	18	26	40	53	69	109	157
	4	1	1	3	6	8	13	19	30	40	51	81	117
	3	1	1	3	5	7	11	16	25	34	44	69	100
	2	1	1	2	4	6	10	14	22	29	37	59	85
	1	0	1	1	3	4	7	10	15	20	26	41	60
	1/0	0	1	1	1	3	6	8	13	17	72	35	51
	2/0	0	1	1	1	3	5	7	11	15	19	30	43
	3/0	0	1	1	1	2	4	6	9	12	1b	25	3b
	4/0	0	0	1	1	1	3	5	8	10	13	21	30
	250	0	0	1	1	1	3	4	6	8	11	17	25
	300	0	0	1	1	1	2	3	5	7	9	15	21
	350	0	0	0	1	1	1	3	5	6	8	13	19
	400	0	0	0	1	1	1	3	4	6	7	12	1/
	500	0	0	0	1	1	1	2	3	5	6	10	14
	600	0	0	0	0	1	1	1	3	4	5	8	11
	700	0	0	0	0	1	1	1	2	3	4	7	10
	750	0	0	0	0	1	1	1	2	3	4	6	10
THHN. THWN, THWN 2	14	11	21	34	60	82	135	193	299	401	517	815	1178
	12	8	15	25	43	59	94	141	218	293	3//	594	859
	10	5	9	15	27	37	62	89	137	184	238	374	541
	8	3	5	9	16	21	36	51	/9	106	13/	21b	312
	6	1	4	6	11	15	2b	37	57	77	99	156	225
	4	1	2	4	7	9	16	22	36	47	61	96	138
	3	1	1	3	6	8	13	19	30	40	61	81	117
	2	1	1	3	5	7	11	16	25	33	43	68	98
	1	1	1	1	3	5	8	12	18	25	32	50	73

(continued on next page)

MAXIMUM NUMBER OF CONDUCTORS IN RIGID PVC CONDUIT, SCHEDULE 40

Type Letters	Cond.Size AWG/kcmil	½	¾	1	1¼	1½	2	2½	3	3½	4	5	6
THHN, THWN, THWN-2	1/0	1	1	1	3	4	7	10	15	21	27	42	61
	2/0	0	1	1	2	3	6	8	13	17	22	35	51
	3/0	0	1	1	1	3	5	7	11	14	18	29	42
	4/0	0	1	1	1	2	4	6	9	12	15	24	35
	250	0	1	1	1	1	3	4	7	10	12	20	28
	300	0	0	1	1	1	3	3	6	8	11	17	24
	350	0	0	1	1	1	3	3	5	7	9	15	21
	400	0	0	1	1	1	2	3	5	6	8	13	19
	500	0	0	0	1	1	1	3	4	5	7	11	16
	600	0	0	0	1	1	1	1	3	4	5	9	13
	700	0	0	0	0	1	1	1	3	4	5	8	11
	750	0	0	0	0	1	1	1	2	3	4	7	11
FEP, FEPB, PFA, PFAH, TFE	14	11	20	33	58	79	131	188	290	389	502	790	1142
	12	8	15	24	42	50	90	137	212	284	366	577	834
	10	6	10	17	30	41	69	98	152	204	263	414	898
	8	3	8	10	17	24	39	58	87	117	150	237	343
	6	2	4	7	12	17	28	40	62	83	107	169	244
	4	1	3	5	8	12	19	28	43	58	75	118	170
	3	1	2	4	7	10	16	23	36	49	62	98	142
	2	1		3	5	8	13	19	30	40	51	81	117
PFA, PFAH, TFE	1	1	1	2	4	5	9	13	20	28	36	56	81
PFA, PFAH, TFE, Z	1/0	1	1	1	3	4	6	11	17	23	30	47	68
	2/0	0	1	1	3	4	8	9	14	19	24	39	56
	3/0	0	1	1	2	3	8	7	12	18	20	32	46
	4/0	0	1	1	1	2	4	8	8	14	16	26	38
Z	14	13	24	40	70	95	158	226	350	469	605	952	1376
	12	9	17	28	49	68	112	160	248	333	429	675	976
	10	6	10	17	30	41	69	98	152	204	263	414	598
	8	3	6	11	19	28	43	62	96	129	166	261	378
	6	2	4	7	14	19	30	43	67	90	116	184	265
	4	1	3	5	9	13	21	30	46	62	80	126	183
	3	1	2	4	6	9	15	22	34	45	58	92	133
	2	1	1	3	5	7	12	18	28	38	49	77	111
	1	1	1	2	4	6	10	14	23	30	39	62	90
XHH, XHHW, XHHW-2, ZW	14	6	14	24	42	57	94	135	209	280	361	566	822
	12	6	11	18	32	44	72	103	160	215	277	436	631
	10	4	8	13	24	32	54	77	119	160	206	325	470
	8	2	4	7	13	18	30	43	68	89	115	181	261
	6	1	3	5	10	13	22	32	49	66	85	134	193
	4	1	2	4	7	9	16	23	35	48	61	97	140
	3	1	1	3	6	8	13	19	30	41	52	82	118
	2	1	1	3	5	7	11	16	25	34	44	69	99
XHH, XHHW, XHHW-2	1	1	1	1	3	5	8	12	19	25	32	51	74
	1/0	1	1	1	3	4	7	10	16	21	27	43	62
	2/0	0	1	1	2	3	6	8	13	18	23	36	52
	3/0	0	1	1	2	3	5	7	11	14	19	30	43
	4/0	0	1	1	1	2	4	6	9	12	16	24	35
	250	0	0	1	1	1	3	5	7	10	12	20	29
	300	0	0	1	1	1	3	4	6	8	11	17	25
	350	0	0	1	1	1	2	3	5	7	9	15	22
	400	0	0	0	1	1	1	3	5	6	8	13	19
	500	0	0	0	1	1	1	2	4	5	7	11	16
	600	0	0	0	1	1	1	1	3	4	5	9	13
	700	0	0	0	0	1	1	1	3	4	5	8	11
	750	0	0	0	0	1	1	1	2	3	4	7	11

*Types RHH, RHW, and RHW-2 without outer covering.

See *Ugly's* page 141 for Trade Size/Metric Designator conversion.

Source: NFPA 70®, *National Electrical Code*®, 2023 edition, NFPA, Quincy, MA, 2022, Annex C, Table C.11, as modified.

MAXIMUM NUMBER OF CONDUCTORS IN RIGID PVC CONDUIT, SCHEDULE 80

Type Letters	Cond.Size AWG/kcmil	Trade Sizes in Inches											
		½	¾	1	1¼	1½	2	2½	3	3½	4	5	6
RHH, RHW, RHW-2	14	3	5	9	17	23	39	56	88	118	153	243	349
	12	2	4	7	14	19	32	46	73	98	127	202	290
	10	1	3	6	11	15	26	37	59	79	103	163	234
	8	1	1	3	6	8	13	19	31	41	54	85	122
	6	1	1	2	4	6	11	16	24	33	43	68	98
	4	1	1	1	3	5	8	12	19	26	33	53	77
	3	1	1	1	3	4	7	11	17	23	29	47	67
	2	0	1	1	3	4	6	9	14	20	25	41	58
	1	0	1	1	1	2	4	6	9	13	17	27	38
	1/0	0	0	1	1	1	3	5	8	11	15	23	33
	2/0	0	0	1	1	1	3	4	7	10	13	20	29
	3/0	0	0	1	1	1	3	4	6	8	11	17	25
	4/0	0	0	0	1	1	2	3	5	7	9	15	21
	250	0	0	0	1	1	1	2	4	5	7	11	16
	300	0	0	0	0	1	1	2	3	5	6	10	14
	350	0	0	0	1	1	1	1	3	4	5	9	13
	400	0	0	0	0	1	1	1	3	4	5	8	12
	500	0	0	0	0	0	1	1	1	3	3	6	8
	600	0	0	0	0	0	1	1	1	3	3	6	8
	700	0	0	0	0	0	1	1	1	2	3	5	7
	750	0	0	0	0	0	0	1	1	2	3	5	7
TW, THHW, THW, THW-2	14	6	11	19	35	49	82	118	185	250	324	514	736
	12	4	9	15	27	38	63	91	142	192	248	394	565
	10	3	6	11	20	28	47	68	106	143	185	294	421
	8	1	3	6	11	15	26	37	59	79	103	163	234
RHH*. RHW.* RHW-2*	14	4	8	13	23	32	55	79	123	166	215	341	490
	12	3	6	10	19	26	44	63	99	133	173	274	394
	10	2	5	8	15	20	34	49	77	104	135	214	307
	8	1	3	5	9	12	20	29	46	62	81	128	184
RHH*. RHW.* RHW-2*. TW. THW. THHW. THW-2	6	1	1	3	7	9	16	22	35	48	62	98	141
	4	1	1	3	5	7	12	17	26	35	46	73	105
	3	1	1	2	4	6	10	14	22	30	39	63	90
	2	1	1	1	3	5	8	12	19	26	33	53	77
	1	0	1	1	2	3	6	8	13	18	23	37	54
	1/0	0	1	1	1	3	5	7	11	15	20	32	46
	2/0	0	1	1	1	2	4	6	10	13	17	27	39
	3/0	0	0	1	1	1	3	5	8	11	14	23	33
	4/0	0	0	1	1	1	3	4	7	9	12	19	27
	250	0	0	0	1	1	2	3	5	7	9	15	22
	300	0	0	0	1	1	1	3	5	6	8	13	19
	350	0	0	0	1	1	1	2	4	5	7	12	18
	400	0	0	0	0	1	1	1	4	5	7	10	15
	500	0	0	0	0	1	1	1	3	4	5	9	13
	600	0	0	0	0	0	1	1	2	3	4	7	10
	700	0	0	0	0	0	1	1	1	3	4	6	9
	750	0	0	0	0	0	1	1	1	3	4	6	8
THHN, THWN, THWN 2	14	9	17	28	51	70	118	170	265	358	464	736	1055
	12	6	12	20	37	51	86	124	193	261	338	537	770
	10	4	7	13	23	32	54	78	122	164	213	338	485
	8	2	4	7	13	18	31	45	70	95	123	195	279
	6	1	3	5	9	13	22	32	51	68	89	141	202
	4	1	1	3	6	8	14	20	31	42	54	86	124
	3	1	1	3	5	7	12	17	26	35	46	73	105
	2	1	1	2	4	6	10	14	22	30	39	61	88
	1	0	1	1	3	4	7	10	16	22	29	45	65

(continued on next page)

MAXIMUM NUMBER OF CONDUCTORS IN RIGID PVC CONDUIT, SCHEDULE 80

Type Letters	Cond.Size AWG/kcmil	½	¾	1	1¼	1½	2	2½	3	3½	4	5	6
THHN, THWN, THWN-2	1/0	0	1	1	1	3	6	9	14	18	24	38	55
	2/0	0	1	1	1	3	5	7	11	15	20	32	48
	3/0	0	1	1	1	2	4	6	9	13	17	26	38
	4/0	0	0	1	1	1	3	5	8	10	14	22	31
	250	0	0	1	1	1	3	4	6	8	11	18	25
	300	0	0	0	1	1	2	3	5	7	9	15	22
	350	0	0	0	1	1	1	3	5	6	8	13	19
	400	0	0	0	1	1	1	3	4	6	7	12	17
	500	0	0	0	0	1	1	2	3	5	6	10	14
	600	0	0	0	0	1	1	1	3	4	5	8	12
	700	0	0	0	0	0	1	1	2	3	4	7	10
	750	0	0	0	0	0	1	1	2	3	4	7	9
FEP, FEPB, PFA, PFAH, TFE	14	8	16	27	49	68	115	164	257	347	450	714	1024
	12	6	12	20	36	50	84	120	188	253	328	521	747
	10	4	8	14	26	36	60	86	135	182	235	374	536
	8	2	5	8	15	20	34	49	77	104	135	214	307
	6	1	3	6	10	14	24	35	55	74	96	152	218
	4	1	2	4	7	10	17	24	38	62	87	106	153
	3	1	1	3	6	8	14	20	37	43	56	99	127
	2	1	1	3	5	7	12	17	26	35	48	73	105
PFA, PFAH, TFE	1	1	1	1	1	3	5	8	11	18	25	51	73
PFA, PFAH, TFE, Z	1/0	0	1	1	1	4	7	10	15	20	27	42	61
	2/0	0	1	1	1	3	5	8	12	17	22	35	50
	3/0	0	1	1	1	2	4	6	10	14	18	29	41
	4/0	0	0	1	1	1	4	5	8	11	15	24	34
Z	14	10	19	33	59	82	138	198	310	418	542	860	1233
	12	7	14	23	42	58	98	141	220	297	385	610	875
	10	4	8	14	26	36	60	86	135	182	235	374	536
	8	3	5	9	16	22	38	54	86	115	149	236	339
	6	1	4	6	11	16	26	38	60	81	104	166	238
	4	1	2	4	8	11	18	26	41	55	72	114	164
	3	1	1	3	5	8	13	19	30	40	52	83	119
	2	1	1	2	5	6	11	16	25	33	43	69	99
	1	1	1	1	4	4	9	13	20	27	35	56	80
XHH, XHHW, XHHW-2, ZW	14	6	11	19	35	49	82	118	185	250	324	514	736
	12	4	9	15	27	38	63	91	142	192	248	394	565
	10	3	6	11	20	28	47	68	106	143	185	294	421
	8	1	3	6	11	15	26	37	59	79	103	163	234
	6	1	2	4	8	11	19	28	43	59	76	121	173
	4	1	1	3	6	8	14	20	31	42	55	87	125
	3	1	1	3	5	7	12	17	26	36	47	74	106
	2	1	1	2	4	6	10	14	22	30	39	62	89
XHH, XHHW, XHHW-2	1	0	1	1	3	4	7	10	16	22	29	46	66
	1/0	0	1	1	2	3	6	9	14	19	24	39	56
	2/0	0	1	1	1	3	5	7	11	16	20	32	46
	3/0	0	1	1	1	2	4	6	9	13	17	27	38
	4/0	0	0	1	1	2	4	6	8	11	14	22	32
	250	0	0	1	1	1	3	4	6	9	11	18	26
	300	0	0	1	1	1	2	3	5	7	10	15	22
	350	0	0	0	1	1	1	3	5	6	8	14	20
	400	0	0	0	1	1	1	3	4	6	7	12	17
	500	0	0	0	1	1	1	2	3	5	6	10	14
	600	0	0	0	0	1	1	1	3	4	5	9	11
	700	0	0	0	0	1	1	1	2	3	4	7	10
	750	0	0	0	0	1	1	1	2	3	4	6	9

*Types RHH, RHW, and RHW-2 without outer covering.

See Ugly's page 141 for Trade Size/Metric Designator conversion.

Source: NFPA 70®, *National Electrical Code®*, 2023 edition, NFPA, Quincy, MA, 2022, Annex C, Table C.10, as modified.

MAXIMUM NUMBER OF CONDUCTORS IN RIGID METAL CONDUIT

Type Letters	Cond.Size AWG/kcmil	Trade Sizes in Inches											
		½	¾	1	1¼	1½	2	2½	3	3½	4	5	6
RHH, RHW. RHW-2	14	4	7	12	21	26	46	65	102	136	176	276	395
	12	3	6	10	17	23	38	55	85	113	146	229	330
	10	3	5	8	14	19	31	44	68	91	118	185	267
	8	1	2	4	7	10	16	23	36	48	61	97	139
	6	1	1	3	6	8	13	18	29	38	49	77	112
	4	1	1	2	4	6	10	14	22	30	38	60	87
	3	1	1	1	3	5	8	12	19	26	34	53	76
	2	1	1	1	3	4	7	11	17	23	29	46	68
	1	0	1	1	1	3	5	7	11	15	18	30	44
	1/0	0	1	1	1	2	4	6	10	13	17	26	33
	2/0	0	1	1	1	2	4	5	8	11	14	23	33
	3/0	0	0	1	1	1	3	4	7	10	12	20	26
	4/0	0	0	1	1	1	2	4	6	8	11	17	24
	250	0	0	0	1	1	1	3	4	6	8	13	18
	300	0	0	0	1	1	1	2	4	5	7	11	16
	350	0	0	0	1	1	1	2	4	4	6	10	15
	400	0	0	0	1	1	1	1	3	4	5	9	13
	500	0	0	0	1	1	1	1	3	4	5	8	11
	600	0	0	0	0	1	1	1	2	3	4	6	9
	700	0	0	0	0	1	1	1	1	3	3	6	8
	750	0	0	0	0	0	1	1	1	3	3	5	6
TW. THHN, THW. THW-2	14	9	15	25	44	59	98	140	215	238	370	381	639
	12	7	12	19	33	45	75	107	165	221	284	446	644
	10	5	9	14	25	34	56	80	123	164	212	332	480
	8	3	5	8	14	19	31	44	68	91	118	165	267
RHH*. RHW.* RHW-2*	14	6	10	17	29	39	65	93	143	191	246	367	558
	12	5	8	13	23	32	52	75	115	154	198	311	448
	10	3	6	10	18	25	41	58	90	120	154	242	350
	8	1	4	6	11	15	24	35	54	72	92	145	209
RHH*. RHW.* RHW-2*. TW. THW, THHN, THW-2	6	1	3	5	8	11	18	27	41	55	71	111	160
	4	1	1	3	6	8	14	20	31	41	53	83	120
	3	1	1	2	5	7	12	17	26	35	45	71	103
	2	1	1	1	4	6	10	14	22	30	38	60	87
	1	1	1	1	3	4	7	10	15	21	27	42	61
	1/0	0	1	1	2	3	6	8	13	18	23	36	52
	2/0	0	1	1	2	3	5	7	11	15	19	31	44
	3/0	0	1	1	1	2	4	6	9	13	16	26	37
	4/0	0	0	1	1	1	3	5	8	10	14	21	31
	250	0	0	1	1	1	3	4	6	8	11	17	25
	300	0	0	1	1	1	2	3	5	7	9	15	22
	350	0	0	0	1	1	1	3	5	6	8	13	19
	400	0	0	0	1	1	1	3	4	6	7	12	17
	500	0	0	0	1	1	1	2	3	5	6	10	14
	600	0	0	0	1	1	1	1	3	4	5	8	12
	700	0	0	0	0	1	1	1	2	3	4	7	10
	750	0	0	0	0	1	1	1	2	3	4	7	10
THHN. THWN, THWN-2	14	13	22	36	63	85	140	200	309	412	531	833	1202
	12	9	16	26	46	62	102	146	225	301	387	608	877
	10	6	10	17	29	39	64	92	142	189	244	383	552
	8	3	6	9	16	22	37	53	82	109	140	221	318
	6	2	4	7	12	16	27	38	59	79	101	159	230
	4	1	2	4	7	10	16	23	36	48	62	98	141
	3	1	1	3	6	8	14	20	31	41	53	83	120
	2	1	1	3	5	7	11	17	26	34	44	70	100
	1	1	1	1	4	5	8	12	19	25	33	51	74

(continued on next page)

MAXIMUM NUMBER OF CONDUCTORS IN RIGID METAL CONDUIT

Type Letters	Cond. Size AWG/kcmil	Trade Sizes in Inches											
		½	¾	1	1¼	1½	2	2½	3	3½	4	5	6
THHN, THWN, THWN-2	1/0	1	1	1	3	4	7	10	16	21	27	43	63
	2/0	0	1	1	2	3	6	8	13	18	23	36	52
	3/0	0	1	1	1	3	5	7	11	15	19	30	43
	4/0	0	1	1	1	2	4	6	9	12	16	25	36
	250	0	0	1	1	1	3	5	7	10	13	20	29
	300	0	0	1	1	1	3	4	6	8	11	17	25
	350	0	0	1	1	1	3	4	6	8	10	15	22
	400	0	0	1	1	1	2	3	5	7	8	13	20
	500	0	0	0	1	1	2	3	4	5	7	11	16
	600	0	0	0	1	1	1	2	3	4	6	9	13
	700	0	0	0	1	1	1	1	3	4	5	8	11
	750	0	0	0	0	1	1	1	3	4	5	7	11
FEP, FEPB, PFA, PFAH, TFE	14	12	22	35	61	83	136	194	300	400	515	808	1166
	12	9	16	26	44	60	99	142	219	292	376	590	851
	10	6	11	18	32	43	71	102	157	209	269	423	610
	8	3	6	10	18	25	41	58	90	120	154	242	350
	6	2	4	7	13	17	29	41	64	85	110	172	249
	4	1	3	5	9	12	20	29	44	59	77	120	174
	3	1	2	4	7	10	17	24	37	50	64	100	145
	2	1	1	3	6	8	14	20	31	41	53	83	120
PFA, PFAH, TFE	1	1	1	2	4	6	9	14	21	28	37	57	83
PFA, PFAH, TFE, Z	1/0	1	1	1	3	5	8	11	18	24	30	48	69
	2/0	0	1	1	3	4	6	9	14	19	25	40	57
	3/0	0	1	1	2	3	5	8	12	16	21	33	47
	4/0	0	1	1	1	2	4	6	10	13	17	27	39
Z	14	15	26	42	73	100	164	234	361	482	621	974	1405
	12	10	18	30	52	71	116	166	256	342	440	691	997
	10	6	11	18	32	43	71	102	157	209	269	423	610
	8	4	7	11	20	27	45	64	99	132	170	267	386
	6	3	5	8	14	19	31	45	69	93	120	188	271
	4	1	3	5	9	13	22	31	48	64	82	129	186
	3	1	2	4	7	9	16	22	35	47	60	94	136
	2	1	1	3	6	8	13	19	29	39	50	78	113
	1	1	1	2	5	6	10	15	23	31	40	63	92
XHH, XHHW, XHHW-2, ZW	14	9	15	25	44	59	98	140	215	288	370	581	839
	12	7	12	19	33	45	75	107	165	221	284	446	644
	10	5	9	14	25	34	56	80	123	164	212	332	480
	8	3	5	8	14	19	31	44	68	91	118	185	267
	6	1	3	6	10	14	23	33	51	68	87	137	197
	4	1	2	4	7	10	16	24	37	49	63	99	143
	3	1	1	3	6	8	14	20	31	41	53	84	121
	2	1	1	3	5	7	12	17	26	35	45	70	101
XHH, XHHW, XHHW-2	1	1	1	1	4	5	9	12	19	26	33	52	76
	1/0	1	1	1	3	4	7	10	16	22	28	44	64
	2/0	0	1	1	2	3	6	9	13	18	23	37	53
	3/0	0	1	1	1	3	5	7	11	15	19	30	44
	4/0	0	1	1	1	2	4	6	9	12	16	25	36
	250	0	0	1	1	1	3	5	7	10	13	20	30
	300	0	0	1	1	1	3	4	6	9	11	18	25
	350	0	0	1	1	1	2	3	6	7	10	15	22
	400	0	0	1	1	1	2	3	5	7	9	14	20
	500	0	0	0	1	1	1	2	4	5	7	11	16
	600	0	0	0	1	1	1	1	3	4	6	9	13
	700	0	0	0	1	1	1	1	3	4	5	8	11
	750	0	0	0	0	1	1	1	3	4	5	7	11

*Types RHH, RHW, and RHW-2 without outer covering.

See Ugly's page 141 for Trade Size/Metric Designator conversion.

*Source: NFPA 70®, National Electrical Code®, 2023 edition, NFPA, Quincy, MA, 2022, Annex C, Table C.9, as modified.

MAXIMUM NUMBER OF CONDUCTORS IN FLEXIBLE METAL CONDUIT

Type Letters	Cond.Size AWG/kcmil	Trade Sizes in Inches										
		3/8	1/2	3/4	1	1¼	1½	2	2½	3	3½	4
RHH, RHW, RHW-2	14	1	4	7	11	17	25	44	67	96	131	171
	12	1	3	6	9	14	21	37	55	80	109	142
	10	1	3	5	7	11	17	30	45	64	88	115
	8	0	1	2	4	6	9	15	23	34	46	60
	6	0	1	1	3	5	7	12	19	27	37	48
	4	0	1	1	2	4	5	10	14	21	29	37
	3	0	1	1	1	3	5	8	13	18	25	33
	2	0	1	1	1	3	4	7	11	16	22	28
	1	0	0	1	1	1	2	5	7	10	14	19
	1/0	0	0	1	1	1	2	4	6	9	12	16
	2/0	0	0	1	1	1	1	3	5	8	11	14
	3/0	0	0	0	1	1	1	3	5	7	9	12
	4/0	0	0	0	1	1	1	2	4	6	8	10
	250	0	0	0	0	1	1	1	3	4	6	8
	300	0	0	0	0	1	1	1	2	4	5	7
	350	0	0	0	0	1	1	1	2	3	5	6
	400	0	0	0	0	1	1	1	1	3	4	6
	500	0	0	0	0	0	1	1	1	3	3	5
	600	0	0	0	0	0	1	1	1	2	3	4
	700	0	0	0	0	0	0	1	1	1	3	3
	750	0	0	0	0	0	0	1	1	1	2	3
TW, THHW, THW, THW-2	14	3	9	15	23	36	53	94	141	203	277	361
	12	2	7	11	18	28	41	72	108	156	212	277
	10	1	5	8	13	21	30	54	81	116	158	207
	8	1	3	5	7	11	17	30	45	64	88	115
RHH*, RHW*, RHW-2*	14	1	6	10	15	24	35	62	94	135	184	240
	12	1	5	8	12	19	28	50	75	108	148	193
	10	1	4	6	10	15	22	39	59	85	115	151
	8	1	1	4	6	9	13	23	35	51	69	90
RHH*, RHW*, RHW-2*, TW, THHW, THW-2	6	1	1	3	4	7	10	18	27	39	53	69
	4	0	1	1	3	5	7	13	20	29	39	51
	3	0	1	1	3	4	6	11	17	25	34	44
	2	0	1	1	2	4	5	10	14	21	29	37
	1	0	1	1	1	2	4	7	10	15	20	26
	1/0	0	0	1	1	1	3	6	9	12	17	22
	2/0	0	0	1	1	1	2	5	7	10	14	19
	3/0	0	0	1	1	1	2	4	6	9	12	16
	4/0	0	0	0	1	1	1	3	5	7	10	13
	250	0	0	0	1	1	1	3	4	6	8	11
	300	0	0	0	1	1	1	2	3	5	7	9
	350	0	0	0	1	1	1	1	3	4	6	8
	400	0	0	0	0	1	1	1	3	4	6	7
	500	0	0	0	0	1	1	1	2	3	5	6
	600	0	0	0	0	0	1	1	1	3	4	5
	700	0	0	0	0	0	1	1	1	2	3	4
	750	0	0	0	0	0	0	1	1	1	3	4
THHN, THWN, THWN-2	14	4	13	22	33	52	76	134	202	291	396	518
	12	3	9	16	24	38	56	98	147	212	289	378
	10	1	6	10	15	24	35	62	93	134	182	238
	8	1	3	6	9	14	20	35	53	77	105	137
	6	1	2	4	6	10	14	25	38	55	76	99
	4	0	1	2	4	6	9	16	24	34	46	61
	3	0	1	1	3	5	7	13	20	29	39	51
	2	0	1	1	3	4	6	11	17	24	33	43
	1	0	1	1	1	3	4	8	12	18	24	32

(continued on next page)

94

MAXIMUM NUMBER OF CONDUCTORS IN FLEXIBLE METAL CONDUIT

Type Letters	Cond.Size AWG/kcmil	Trade Sizes in Inches										
		3/8	1/2	3/4	1	1¼	1½	2	2½	3	3½	4
THHN, THWN, THWN-2	1/0	0	1	1	1	2	4	7	10	15	20	27
	2/0	0	0	1	1	1	3	6	9	12	17	22
	3/0	0	0	1	1	1	2	5	7	10	14	18
	4/0	0	0	1	1	1	1	4	6	8	12	15
	250	0	0	0	1	1	1	3	5	7	9	12
	300	0	0	0	1	1	1	3	4	6	8	11
	350	0	0	0	1	1	1	2	3	5	7	9
	400	0	0	0	0	1	1	1	3	5	6	8
	500	0	0	0	0	1	1	1	2	4	5	7
	600	0	0	0	0	1	1	1	1	3	4	5
	700	0	0	0	0	0	1	1	1	3	4	5
	750	0	0	0	0	0	1	1	1	3	4	4
FEP, FEPB, PFA, PFAH, TFE	14	4	12	21	32	51	74	130	196	282	385	502
	12	3	9	15	24	37	54	95	143	206	281	367
	10	2	6	11	17	26	39	68	103	148	201	263
	8	1	4	6	10	15	22	39	59	85	115	151
	6	1	2	4	7	11	16	28	42	60	82	107
	4	1	1	3	5	7	11	19	29	42	57	75
	3	0	1	2	4	6	9	16	24	35	48	62
	2	0	1	1	3	5	7	13	20	29	39	51
PFA, PFAH, TFE	1	0	1	1	2	3	5	9	14	20	27	36
PFA, PFAH, TFE, Z	1/0	0	1	1	1	3	4	8	11	17	23	30
	2/0	0	1	1	1	2	3	6	9	14	19	24
	3/0	0	0	1	1	1	3	5	8	11	15	20
	4/0	0	0	1	1	1	2	4	6	9	13	16
Z	14	5	15	25	39	61	89	157	236	340	463	605
	12	4	11	18	28	43	63	111	168	241	329	429
	10	2	6	11	17	26	39	68	103	148	201	263
	8	1	4	7	11	17	24	43	65	93	127	166
	6	1	3	5	7	12	17	30	45	65	89	117
	4	1	1	3	5	8	12	21	31	45	61	80
	3	0	1	2	4	6	8	15	23	33	45	58
	2	0	1	1	3	5	7	12	19	27	37	49
	1	0	1	1	2	4	6	10	15	22	30	39
XHH, XHHW, XHHW-2, ZW	14	3	9	15	23	36	53	94	141	203	277	361
	12	2	7	11	18	28	41	72	108	156	212	277
	10	1	5	8	13	21	30	54	81	116	158	207
	8	1	3	5	7	11	17	30	45	64	88	115
	6	1	1	3	5	8	12	22	33	48	65	85
	4	0	1	2	4	6	9	16	24	34	47	61
	3	0	1	1	3	5	7	13	20	29	40	52
	2	0	1	1	3	4	6	11	17	24	33	44
XHH, XHHW, XHHW-2	1	0	1	1	1	3	5	8	13	18	25	32
	1/0	0	1	1	1	2	4	7	10	15	21	27
	2/0	0	0	1	1	2	3	6	9	13	17	23
	3/0	0	0	1	1	1	3	5	7	10	14	19
	4/0	0	0	1	1	1	2	4	6	9	12	15
	250	0	0	0	1	1	1	3	5	7	10	12
	300	0	0	0	1	1	1	3	4	6	8	11
	350	0	0	0	1	1	1	2	4	5	7	9
	400	0	0	0	0	1	1	1	3	5	6	8
	500	0	0	0	0	1	1	1	3	4	5	7
	600	0	0	0	0	0	1	1	1	3	4	5
	700	0	0	0	0	0	1	1	1	3	4	5
	750	0	0	0	0	0	1	1	1	2	3	4

*Types RHH, RHW, and RHW-2 without outer covering.

See *Ugly's* page 141 for Trade Size/Metric Designator conversion.

Source: NFPA 70®, *National Electrical Code*®, 2023 edition, NFPA, Quincy, MA, 2022, Annex C, Table C.3, as modified.

MAXIMUM NUMBER OF CONDUCTORS IN LIQUIDTIGHT FLEXIBLE METAL CONDUIT

Type Letters	Cond.Size AWG/kcmil	Trade Sizes in Inches										
		⅜	½	¾	1	1¼	1½	2	2½	3	3½	4
RHH, RHW, RHW-2	14	2	4	7	12	21	27	44	66	102	133	173
	12	1	3	6	10	17	22	36	55	84	110	144
	10	1	3	5	8	14	18	29	44	68	89	116
	8	1	1	2	4	7	9	15	23	36	46	61
	6	1	1	1	3	6	7	12	18	28	37	48
	4	0	1	1	2	4	6	9	14	22	29	38
	3	0	1	1	1	4	5	8	13	19	25	33
	2	0	1	1	1	3	4	7	11	17	22	29
	1	0	0	1	1	1	3	5	7	11	14	19
	1/0	0	0	1	1	1	2	4	6	10	13	16
	2/0	0	0	1	1	1	1	3	5	8	11	14
	3/0	0	0	0	1	1	1	3	4	7	9	12
	4/0	0	0	0	0	1	1	2	4	6	8	10
	250	0	0	0	0	1	1	1	3	4	6	8
	300	0	0	0	0	1	1	1	2	4	5	7
	350	0	0	0	0	1	1	1	2	3	5	6
	400	0	0	0	0	1	1	1	1	3	4	6
	500	0	0	0	0	0	1	1	1	3	4	5
	600	0	0	0	0	0	1	1	1	2	3	4
	700	0	0	0	0	0	0	1	1	1	3	3
	750	0	0	0	0	0	0	1	1	1	2	3
TW, THHW, THW, THW-2	14	5	9	15	25	44	57	93	140	215	280	365
	12	4	7	12	19	33	43	71	108	165	215	280
	10	3	5	9	14	25	32	53	80	123	160	209
	8	1	3	5	8	14	18	29	44	68	89	116
RHH*, RHW*, RHW-2*	14	3	6	10	16	29	38	62	93	143	186	243
	12	3	5	8	13	23	30	50	75	115	149	195
	10	1	3	6	10	18	23	39	58	89	117	152
	8	1	1	4	6	11	14	23	35	53	70	91
RHH*, RHW*, RHW-2*, TW, THHW, THW, THW-2	6	1	1	3	5	8	11	18	27	41	53	70
	4	1	1	1	3	6	8	13	20	30	40	52
	3	1	1	1	3	5	7	11	17	26	34	44
	2	0	1	1	2	4	6	9	14	22	29	38
	1	0	1	1	1	3	4	7	10	15	20	26
	1/0	0	0	1	1	2	3	6	8	13	17	23
	2/0	0	0	1	1	2	3	5	7	11	15	19
	3/0	0	0	1	1	1	2	4	6	9	12	16
	4/0	0	0	0	1	1	1	3	5	8	10	13
	250	0	0	0	1	1	1	3	4	6	8	11
	300	0	0	0	1	1	1	2	3	5	7	9
	350	0	0	0	0	1	1	1	3	5	6	8
	400	0	0	0	0	1	1	1	3	4	6	7
	500	0	0	0	0	1	1	1	2	3	5	6
	600	0	0	0	0	1	1	1	1	3	4	5
	700	0	0	0	0	0	1	1	1	2	3	4
	750	0	0	0	0	0	1	1	1	2	3	4
THHN, THWN, THWN-2	14	8	13	22	36	63	81	134	201	308	401	523
	12	5	9	16	26	46	59	97	146	225	292	381
	10	3	6	10	16	29	37	61	92	141	184	240
	8	1	3	6	9	16	21	35	53	81	106	138
	6	1	2	4	7	12	15	25	38	59	76	100
	4	1	1	2	4	7	9	15	23	36	47	61
	3	1	1	1	3	6	8	13	20	30	40	52
	2	1	1	1	3	5	7	11	17	26	33	44
	1	0	1	1	1	4	5	8	12	19	25	32

(continued on next page)

MAXIMUM NUMBER OF CONDUCTORS IN LIQUIDTIGHT FLEXIBLE METAL CONDUIT

Type Letters	Cond.Size AWG/kcmil	Trade Sizes in Inches										
		3/8	1/2	3/4	1	1 1/4	1 1/2	2	2 1/2	3	3 1/2	4
THHN, THWN, THWN-2	1/0	0	1	1	1	3	4	7	10	16	21	27
	2/0	0	0	1	1	2	3	6	8	13	17	23
	3/0	0	0	1	1	1	3	5	7	11	14	19
	4/0	0	0	1	1	1	2	4	6	9	12	15
	250	0	0	0	1	1	1	3	5	7	10	12
	300	0	0	0	1	1	1	3	4	6	8	11
	350	0	0	0	0	1	1	2	3	5	7	9
	400	0	0	0	0	1	1	1	3	5	6	8
	500	0	0	0	0	1	1	1	2	4	5	7
	600	0	0	0	0	1	1	1	1	3	4	6
	700	0	0	0	0	0	1	1	1	3	4	5
	750	0	0	0	0	0	1	1	1	3	3	5
FEP, FEPB, PFA, PFAH, TFE	14	7	12	21	35	61	79	130	195	299	389	507
	12	5	9	15	25	44	58	94	142	218	284	370
	10	4	6	11	18	32	41	68	102	156	203	266
	8	1	3	6	10	18	23	39	58	89	117	152
	6	1	2	4	7	13	17	27	41	64	83	108
	4	1	1	3	5	9	12	19	29	44	58	75
	3	1	1	2	4	7	10	16	24	37	48	63
	2	1	1	1	3	6	8	13	20	30	40	52
PFA, PFAH, TFE	1	0	1	1	2	4	5	9	14	21	28	36
PFA, PFAH, TFE, Z	1/0	0	1	1	1	3	4	7	11	18	23	30
	2/0	0	1	1	1	3	4	6	9	14	19	25
	3/0	0	0	1	1	2	3	5	8	12	16	20
	4/0	0	0	1	1	1	2	4	6	10	13	17
Z	14	9	15	26	42	73	95	156	235	360	469	611
	12	6	10	18	30	52	67	111	167	255	332	434
	10	4	6	11	18	32	41	68	102	156	203	266
	8	2	4	7	11	20	26	43	64	99	129	168
	6	1	3	5	8	14	18	30	45	69	90	118
	4	1	1	3	5	9	12	20	31	48	62	81
	3	1	1	2	4	7	9	15	23	35	45	59
	2	1	1	1	3	6	7	12	19	29	38	49
	1	1	1	1	2	5	6	10	15	23	30	40
XHH, XHHW, XHHW-2, ZW	14	5	9	15	25	44	57	93	140	215	280	365
	12	4	7	12	19	33	43	71	108	165	215	280
	10	3	5	9	14	25	32	53	80	123	160	209
	8	1	3	5	8	14	18	29	44	68	89	116
	6	1	1	3	6	10	13	22	33	50	66	86
	4	1	1	2	4	7	9	16	24	36	48	62
	3	1	1	1	3	6	8	13	20	31	40	52
	2	1	1	1	3	5	7	11	17	26	34	44
XHH, XHHW, XHHW-2	1	0	1	1	1	4	5	8	12	19	25	33
	1/0	0	1	1	1	3	4	7	10	16	21	28
	2/0	0	0	1	1	2	3	6	9	13	17	23
	3/0	0	0	1	1	1	3	5	7	11	14	19
	4/0	0	0	1	1	1	2	4	6	9	12	16
	250	0	0	0	1	1	1	3	5	7	10	13
	300	0	0	0	1	1	1	3	4	6	8	11
	350	0	0	0	0	1	1	2	3	5	7	10
	400	0	0	0	0	1	1	1	3	5	6	8
	500	0	0	0	0	1	1	1	2	4	5	7
	600	0	0	0	0	1	1	1	1	3	4	6
	700	0	0	0	0	0	1	1	1	3	4	5
	750	0	0	0	0	0	1	1	1	3	3	5

*Types RHH, RHW, and RHW-2 without outer covering.
See *Ugly's* page 141 for Trade Size/Metric Designator conversion.
Source: NFPA 70®, *National Electrical Code®*, 2023 edition, NFPA, Quincy, MA, 2022, Annex C, Table C.8, as modified.

DIMENSIONS OF INSULATED CONDUCTORS AND FIXTURE WIRES

Type	Size	Approx. Area Sq. In.
RFH-2	18	0.0145
FFH-2, RFHH-2	16	0.0172
RHW-2, RHH	14	0.0293
RHW	12	0.0353
	10	0.0437
	8	0.0835
	6	0.1041
	4	0.1333
	3	0.1521
	2	0.1750
	1	0.2660
	1/0	0.3039
	2/0	0.3505
	3/0	0.4072
	4/0	0.4754
	250	0.6291
	300	0.7088
	350	0.7870
	400	0.8626
	500	1.0082
	600	1.2135
	700	1.3561
	750	1.4272
	800	1.4957
	900	1.6377
	1000	1.7719
	1250	2.3479
	1500	2.6938
	1750	3.0357
	2000	3.3719
SF-2, SFF-2	18	0.0115
	16	0.0139
	14	0.0172
SF-1, SFF-1	18	0.0065
RFH-1, XF, XFF	18	0.0088
TF, TFF, XF, XFF	16	0.0109
TW, XF, XFF,	14	0.0139
THHW, THW, THW-2		
TW, THHW,	12	0.0181
THW, THW-2	10	0.0243
	8	0.0437
RHH*, RHW*, RHW-2*	14	0.0209
RHH*, RHW*, RHW-2*,	12	0.0260
XF, XFF		

Type	Size	Approx. Area Sq. In.
RHH*, RHW*, XF	10	0.0333
RHW-2*, XFF		
RHH*, RHW*, RHW-2*	8	0.0556
TW, THW	6	0.0726
THHW	4	0.0973
THW-2	3	0.1134
RHH*	2	0.1333
RHW*	1	0.1901
RHW-2*	1/0	0.2223
	2/0	0.2624
	3/0	0.3117
	4/0	0.3718
	250	0.4596
	300	0.5281
	350	0.5958
	400	0.6619
	500	0.7901
	600	0.9729
	700	1.1010
	750	1.1652
	800	1.2272
	900	1.3561
	1000	1.4784
	1250	1.8602
	1500	2.1695
	1750	2.4773
	2000	2.7818
TFN	18	0.0055
TFFN	16	0.0072
THHN	14	0.0097
THWN	12	0.0133
THWN-2	10	0.0211
	8	0.0366
	6	0.0507
	4	0.0824
	3	0.0973
	2	0.1158
	1	0.1562
	1/0	0.1855
	2/0	0.2223
	3/0	0.2679
	4/0	0.3237
	250	0.3970
	300	0.4608
	350	0.5242
	400	0.5863
	500	0.7073
	600	0.8676
	700	0.9887

DIMENSIONS OF INSULATED CONDUCTORS AND FIXTURE WIRES

Type	Size	Approx. Area Sq. In.	Type	Size	Approx. Area Sq. In.
THHN THWN THWN-2	750 800 900 1000	1.0496 1.1085 1.2311 1.3478	XHHW XHHW-2 XHH	300 350 400 500 600 700 750 800 900 1000 1250 1500 1750 2000	0.4536 0.5166 0.5782 0.6984 0.8709 0.9923 1.0532 1.1122 1.2351 1.3519 1.7180 2.0157 2.3127 2.6073
PF, PGFF, PGF, PFF, PTF, PAF, PTFF, PAFF	18 16	0.0058 0.0075			
PF, PGFF, PGF, PFF, PTF, PAF, PTFF, PAFF TFE, FEP, PFA FEPB, PFAH	14	0.0100			
TFE, FEP, PFA, FEPB, PFAH	12 10 8 6 4 3 2	0.0137 0.0191 0.0333 0.0468 0.0670 0.0804 0.0973	KF-2 KFF-2	18 16 14 12 10	0.0031 0.0044 0.0064 0.0093 0.0139
TFE, PFAH	1	0.1399			
TFE, PFA, PFAH, Z	1/0 2/0 3/0 4/0	0.1676 0.2027 0.2463 0.3000	KF-1 KFF-1	18 16 14 12 10	0.0026 0.0037 0.0055 0.0083 0.0127
ZF, ZFF	18 16	0.0045 0.0061			
Z, ZF, ZFF	14	0.0083			
Z	12 10 8 6 4 3 2 1	0.0117 0.0191 0.0302 0.0430 0.0625 0.0855 0.1029 0.1269			
XHHW, ZW XHHW-2 XHH	14 12 10 8 6 4 3 2	0.0139 0.0181 0.0243 0.0437 0.0590 0.0814 0.0962 0.1146			
XHHW XHHW-2 XHH	1 1/0 2/0 3/0 4/0 250	0.1534 0.1825 0.2190 0.2642 0.3197 0.3904			

*Types RHH, RHW, and RHW-2 without outer covering
See *Ugly's* page 139 for conversion of square inches to square millimeters
Source: NFPA 70®, *National Electrical Code®*, 2023 edition, NFPA, Quincy, MA, 2022, Chapter 9, Table 5, as modified.

COMPACT (STRANDED TYPE) COPPER AND ALUMINUM BUILDING WIRE NOMINAL DIMENSIONS* AND AREAS

Size AWG or kcmil	Bare Conductor		Types THW and THHW		Type THHW		Type XHHW		Size AWG or kcmil
	Number of Strands	Diam. Inches	Approx. Diam. inches	Approx. Area Sq. Inches	Approx. Diam. Inches	Approx. Area Sq. Inches.	Approx. Diam. inches	Approx Area Sq. Inches	
8	8	0.134	0.255	0.0510	—	—	0.224	0.0394	8
6	6	0.169	0.290	0.0660	0.240	0.0452	0.260	0.0530	6
4	4	0.213	0.335	0.0880	0.305	0.0730	0.305	0.0730	4
2	2	0.268	0.390	0.1194	0.360	0.1017	0.360	0.1017	2
1	1	0.299	0.465	0.1698	0.415	0.1352	0.415	0.1352	1
1/0	1/0	0.335	0.500	0.1963	0.450	0.1590	0.450	0.1590	1/0
2/0	2/0	0.376	0.545	0.2332	0.495	0.1924	0.490	0.1885	2/0
3/0	3/0	0.423	0.590	0.2733	0.540	0.2290	0.540	0.2290	3/0
4/0	4/0	0.475	0.645	0.3267	0.595	0.2780	0.590	0.2733	4/0
250	250	0.520	0.725	0.4128	0.670	0.3525	0.660	0.3421	250
300	300	0.570	0.775	0.4717	0.720	0.4071	0.715	0.4015	300
350	350	0.616	0.820	0.5281	0.770	0.4656	0.760	0.4536	350
400	400	0.659	0.865	0.5876	0.815	0.5216	0.800	0.5026	400
500	500	0.736	0.940	0.6939	0.885	0.6151	0.880	0.6082	500
600	600	0.813	1.050	0.8659	0.985	0.7620	0.980	0.7542	600
700	700	0.877	1.110	0.9676	1.050	0.8659	1.050	0.8659	700
750	750	0.908	1.150	1.0386	1.075	0.9076	1.090	0.9331	750
900	900	0.999	1.224	1.1766	1.194	1.1196	1.169	1.0733	900
1000	1000	1.060	1.285	1.2968	1.255	1.2370	1.230	1.1882	1000

*Dimensions are from industry sources.

See *Ugly's* pages 139–140 for metric conversions.

Source: NFPA 70®, *National Electrical Code®*, 2023 edition, NFPA, Quincy, MA, 2022, Chapter 9, Table 5A, as modified.

DIMENSIONS AND PERCENT AREA OF CONDUIT AND TUBING

(For the combinations of wires permitted in Chapter 9, Table 1, *NEC®*)
(See *Ugly's* pages 139-140 for metric conversions.)

Trade Size Inches	Internal Diameter Inches	Total Area 100% SqInches.	2 Wires 31% Sq. Inches	Over 2 Wires 40% Sqj). inches	1 Wires 53% Sq. Inches.	(NIPPLE*) 60% Sq. Inches.
Electrical Metallic Tubing (EMT)							
½	0.622	0.304	0.094	0.122	0.161	0.182
¾	0.824	0.533	0.165	0.213	0.283	0.320
1	1.049	0.864	0.268	0.346	0.458	0.519
1¼	1.380	1.496	0.464	0.598	0.793	0.897
1½	1.610	2.036	0.631	0.814	1.079	1.221
2	2.067	3.356	1.040	1.342	1.778	2.013
2½	2.731	5.858	1.816	2.343	3.105	3.515
3	3.356	8.846	2.742	3.538	4.688	5.307
3½	3.834	11.545	3.579	4.618	6.119	6.927
4	4.334	14.753	4.573	5.901	7.819	8.852
5	5.073	20.212	6.266	8.085	10.713	12.127
6	6.093	29.158	9.039	11.663	15.454	17.495
Electrical Nonmetallic Tubing (ENT)							
½	0.602	0.285	0.088	0.114	0.151	0.171
¾	0.804	0.508	0.157	0.203	0.269	0.305
1	1.029	0.832	0.258	0.333	0.441	0.499
1¼	1.36	1.453	0.450	0.581	0.770	0.872
1½	1.59	1.986	0.616	0.794	1.052	1.191
2	2.047	3.291	1.020	1.316	1.744	1.975
2½	–	–	–	–	–	–
3	–	–	–	–	–	–
3½	–	–	–	–	–	–
4	–	–	–	–	–	–
Flexible Metal Conduit (FMC)							
⅜	0.384	0.116	0.036	0.046	0.061	0.069
½	0.635	0.317	0.098	0.127	0.168	0.190
¾	0.824	0.533	0.165	0.213	0.283	0.320
1	1.020	0.817	0.253	0.327	0.433	0.490
1¼	1.275	1.277	0.396	0.511	0.677	0.766
1½	1.538	1.858	0.576	0.743	0.985	1.115
2	2.040	3.269	1.013	1.307	1.732	1.961
2½	2.500	4.909	1.522	1.963	2.602	2.945
3	3.000	7.069	2.191	2.827	3.746	4.241
3½	3.500	9.621	2.983	3.848	5.099	5.773
4	4.000	12.566	3.896	5.027	6.660	7.540
Intermediate Metal Conduit (IMC)							
⅜	–	–	–	–	–	–
½	0.660	0.342	0.106	0.137	0.181	0.205
¾	0.864	0.586	0.182	0.235	0.311	0.352
1	1.105	0.959	0.297	0.384	0.508	0.575
1¼	1.448	1.647	0.510	0.659	0.873	0.988
1½	1.683	2.225	0.690	0.890	1.179	1.335
2	2.150	3.630	1.125	1.452	1.924	2.178
2½	2.557	5.135	1.592	2.054	2.722	3.081
3	3.176	7.922	2.456	3.169	4.199	4.753
3½	3.671	10.584	3.281	4.234	5.610	6.351
4	4.166	13.631	4.226	5.452	7.224	8.179
5	5.210	21.32	6.610	8.528	11.30	12.792
6	6.258	30.76	9.536	12.304	16.302	18.456

(continued on the next page)

DIMENSIONS AND PERCENT AREA OF CONDUIT AND TUBING

(For the combinations of wires permitted in Chapter 9, Table 1, *NEC*®)
(See *Ugly's* pages 139-140 for metric conversions.)

Trade Size Inches	Internal Diameter Inches	Total Area 100% SqInches.	2 Wires 31% Sq. Inches.	Over 2 Wires 40% Sq). inches	1 Wires 53% Sq. Inches.	(NIPPLE) 60% Sq. Inches.
Liquidting Flexible Nonmetallic Conduit (LFNC-A*)						
⅜	0.495	0.192	0.060	0.077	0.102	0.115
½	0.630	0.312	0.097	0.125	0.165	0.187
¾	0.825	0.535	0.166	0.214	0.283	0.321
1	1.043	0.854	0.265	0.342	0.453	0.513
1¼	1.383	1.502	0.466	0.601	0.796	0.901
1½	1.603	2.018	0.626	0.807	1.070	1.211
2	2.063	3.343	1.036	1.337	1.772	2.006
*Corresponds to 356.2(1).						
Liquidting Flexible Nonmetallic Conduit (LFNC-B*)						
⅜	0.494	0.192	0.059	0.077	0.102	0.115
½	0.632	0.314	0.097	0.125	0.166	0.188
¾	0.830	0.541	0.168	0.216	0.287	0.325
1	1.054	0.873	0.270	0.349	0.462	0.524
1¼	1.395	1.528	0.474	0.611	0.810	0.917
1½	1.588	1.981	0.614	0.792	1.050	1.188
2	2.033	3.246	1.006	1.298	1.720	1.948
*Corresponds to 356.2(2).						
Liquidting Flexible Metal Conduit (LFNC)						
⅜	0.494	0.192	0.059	0.077	0.102	0.115
½	0.632	0.314	0.097	0.125	0.166	0.188
¾	0.830	0.541	0.168	0.216	0.287	0.325
1	1.054	0.873	0.270	0.349	0.462	0.524
1¼	1.395	1.528	0.474	0.611	0.810	0.917
1½	1.588	1.981	0.614	0.792	1.050	1.188
2	2.033	3.246	1.006	1.298	1.720	1.948
2½	2.493	4.881	1.513	1.953	2.587	2.929
3	3.085	7.475	2.317	2.990	3.962	4.485
3½	3.520	9.731	3.017	3.893	5.158	5.839
4	4.020	12.692	3.935	5.077	6.727	7.615
Rigid Metal Conduit (RMC)						
⅜	–	–	–	–	–	–
½	0.632	0.314	0.097	0.125	0.166	0.188
¾	0.836	0.549	0.170	0.220	0.291	0.329
1	1.063	0.887	0.275	0.355	0.470	0.532
1¼	1.394	1.526	0.473	0.610	0.809	0.916
1½	1.624	2.071	0.642	0.829	1.098	1.243
2	2.083	3.408	1.056	1.363	1.806	2.045
2½	2.489	4.866	1.508	1.946	2.579	2.919
3	3.090	7.499	2.325	3.000	3.974	4.499
3½	3.570	10.010	3.103	4.004	5.305	6.006
4	4.050	12.882	3.994	5.153	6.828	7.729
5	5.073	20.212	6.266	8.085	10.713	12.127
6	6.093	29.158	9.039	11.663	15.454	17.495

DIMENSIONS AND PERCENT AREA OF CONDUIT AND TUBING

(For the combinations of wires permitted in Table 1, Chapter 9, NEC®)
(See *Ugly's* pages 139-140 for metric conversions.)

Trade Size Inches	Internal Diameter Inches	Total Area 100% SqInches.	2 Wires 31% Sq. Inches	Over 2 Wires 40% Sq), inches	1 Wires 53% Sq. Inches.	(NIPPLE) 60% Sq. Inches
Rigid PVC Conduit (PVC), Schedule 80						
½	0.526	0.217	0.067	0.087	0.115	0.130
¾	0.722	0.409	0.127	0.164	0.217	0.246
1	0.936	0.688	0.213	0.275	0.365	0.413
1¼	1.255	1.237	0.383	0.495	0.656	0.742
1½	1.476	1.711	0.530	0.684	0.907	1.027
2	1.913	2.874	0.891	1.150	1.523	1.725
2½	2.290	4.119	1.277	1.647	2.183	2.471
3	2.864	6.442	1.997	2.577	3.414	3.865
3½	3.326	8.688	2.693	3.475	4.605	5.213
4	3.786	11.258	3.490	4.503	5.967	6.755
5	4.768	17.855	5.535	7.142	9.463	10.713
6	5.709	25.598	7.935	10.239	13.567	15.359
Rigid PVC Conduit (PVC), Schedule 40 & HDPE Conduit (HDPE)						
½	0.602	0.285	0.088	0.114	0.151	0.171
¾	0.804	0.508	0.157	0.203	0.269	0.305
1	1.029	0.832	0.258	0.333	0.441	0.499
1¼	1.360	1.453	0.450	0.581	0.770	0.872
1½	1.590	1.986	0.616	0.794	1.052	1.191
2	2.047	3.291	1.020	1.316	1.744	1.975
2½	2.445	4.695	1.455	1.878	2.488	2.817
3	3.042	7.268	2.253	2.907	3.852	4.361
3½	3.521	9.737	3.018	3.895	5.161	5.842
4	3.998	12.554	3.892	5.022	6.654	7.532
5	5.016	19.761	6.126	7.904	10.473	11.856
6	6.031	28.567	8.856	11.427	15.141	17.140
Type, A, Rigid PVC Conduit (PVC)						
½	0.700	0.385	0.119	0.154	0.204	0.231
¾	0.910	0.650	0.202	0.260	0.345	0.390
1	1.175	1.084	0.336	0.434	0.575	0.651
1¼	1.500	1.767	0.548	0.707	0.937	1.060
1½	1.720	2.324	0.720	0.929	1.231	1.394
2	2.155	3.647	1.131	1.459	1.933	2.188
2½	2.635	5.453	1.690	2.181	2.890	3.272
3	3.230	8.194	2.540	3.278	4.343	4.916
3½	3.690	10.694	3.315	4.278	5.668	6.416
4	4.180	13.723	4.254	5.489	7.273	8.234
Type EB, PVC Conduit (PVC)						
2	2.221	3.874	1.201	1.550	2.053	2.325
2½	–	–	–	–	–	–
3	3.330	8.709	2.700	3.484	4.616	5.226
3½	3.804	11.365	3.523	4.546	6.023	6.819
4	4.289	14.448	4.479	5.779	7.657	8.669
5	5.316	22.195	6.881	8.878	11.763	13.317
6	6.336	31.530	9.774	12.612	16.711	18.918

*Nipples must be 24 inches or less.

Source: NFPA 70®, *National Electrical Code®*, 2023 edition, NFPA, Quincy, MA, 2022, Chapter Table 4. as modified.

THREAD DIMENSIONS AND TAP DRILL SIZES

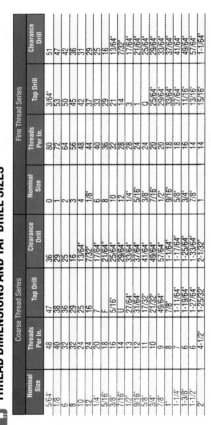

Coarse Thread Series				Fine Thread Series			
Nominal Size	Threads Per In.	Tap Drill	Clearance Drill	Nominal Size	Threads Per In.	Tap Drill	Clearance Drill
5/64"	48	47	36	0	80	3/64"	51
1/8"	40	38	29	1	72	53	47
6	32	36	25	2	64	50	42
8	32	29	16	3	56	45	36
10	24	25	13/64"	4	48	42	31
12	24	16	7/32"	1/8"	44	37	29
1/4"	20	7	17/64"	6	40	33	25
5/16"	18	F	21/64"	8	36	29	16
3/8"	16	5/16	25/64"	10	32	21	13/64"
7/16"	14	U	29/64"	12	28	14	7/32"
1/2"	13	27/64"	33/64"	1/4"	28	3	17/64"
9/16"	12	31/64"	37/64"	5/16"	24	1	21/64"
5/8"	11	17/32	41/64"	3/8"	24	Q	25/64"
3/4"	10	21/32"	49/64"	7/16"	20	25/64"	29/64"
7/8"	9	49/64"	57/64"	1/2"	20	29/64"	33/64"
1"	8	7/8"	1-1/64"	9/16"	18	33/64"	37/64"
1-1/4"	7	1-11/64"	1-7/64"	5/8"	18	37/64"	41/64"
1-3/8"	7	1-19/64"	1-25/64"	3/4"	16	11/16	49/64"
1-1/2"	6	1-27/64"	1-33/64"	7/8"	14	13/16	57/64"
2"	4-1/2	1-25/32"	2-1/32"	1	14	15/16	1-1/64"

HOLE SAW CHART*

Trade Size	Rigid Conduit	E.M.T. Conduit	Green-Field	L.T. Flex
1/2"	7/8"	3/4	1"	1-1/8"
3/4	1-1/8"	1"	1-1/8"	1-1/4"
1"	1-3/8"	1-1/4"	1-1/2"	1-1/2"
1-1/4"	1-3/4"	1-5/8"	1-3/4"	1-7/8"
1-1/2"	2"	1-7/8"	2"	2-1/8"
2"	2-1/2"	2-1/8"	2-1/2"	2-3/4"

Trade Size	Rigid Conduit	E.M.T. Conduit	Green-Field
2-1/2"	3"	2-7/8"	2-7/8"
3"	3-5/8	3-1/2	3-5/8"
3-1/2"	4-1/8"	4	4-1/8"
4"	4-5/8"	4-1/2"	4-5/8"
5"	5-3/4"		
6"	6-3/4"		

Note: For oil-type push button station, use size 1-7/32 knock-out punch.

METAL BOXES

Box Dimension Inches Trade Size or Type	Min. Cu. In. Capacity	Maximum Number of Conductors						
		No. 18	No. 16	No. 14	No. 12	No. 10	No. 8	No. 6
4 x 1-1/4 Round or Octagonal	12.5	8	7	6	5	5	4	2
4 x 1-1/2 Round or Octagonal	15.5	10	8	7	6	6	5	3
4 x 2-1/8 Round or Octagonal	21.5	14	12	10	9	8	7	4
4 x 1-1/4 Square	18.0	12	10	9	8	7	6	3
4 x 1-1/2 Square	21.0	14	12	10	9	8	7	4
4 x 2-1/8 Square	30.3	20	17	15	13	12	10	6
4-11/16 x 1-1/4 Square	25.5	17	14	12	11	10	8	5
4-11/16 x 1-1/2 Square	29.5	19	16	14	13	11	9	5
4-11/16 x 1-1/8 Square	42.0	28	24	21	18	16	14	8
3 x 2 x 1-1/2 Device	7.5	5	4	3	3	3	2	1
3 x 2 x 2 Device	10.0	6	5	5	4	4	3	2
3 x 2 x 2-1/4 Device	10.5	7	6	5	4	4	3	2
3 x 2 x 2-1/2 Device	12.5	8	7	6	5	5	4	2
3 x 2 x 2-3/4 Device	14.0	9	8	7	6	5	4	2
3 x 2 x 3-1/4 Device	18.0	12	10	9	8	7	6	3
4 x 2-1/8 x 1-1/2 Device	10.3	6	5	5	4	4	3	2
4 x 2-1/8 x 1-7/8 Device	13.0	8	7	6	5	5	4	2
4 x 2-1/8 x 2-1/8 Device	14.5	9	8	7	6	5	4	2
3-3/4 x 2 x 2-1/2 Masonry Box/Gang	14.0	9	8	7	6	5	4	2
3-3/4 x 2 x 3-1/2 Masonry Box/Gang	21.0	14	12	10	9	8	7	4
FS Minimum Internal Depth 1-3/4 Single Cover/Gang	13.5	9	7	6	6	5	4	2
FD Minimum Internal Depth 2-3/8 Single Cover/Gang	18.0	12	10	9	8	7	6	3
FS Minimum Internal Depth 1-3/4 Multiple Cover/Gang	18.0	12	10	9	8	7	6	3
FD Minimum Internal Depth 2-3/8 Multiple Cover/Gang	24.0	16	13	12	10	9	8	4

Source: NFPA 70®. *National Electrical Code®*, 2023 edition. NFPA, Quincy, MA. 2022. Table 314.16(A), as modified.

MINIMUM COVER REQUIREMENTS, 0 To 1000 VOLTS ac, 1500 VOLTS dc, NOMINAL

Cover is defined as the distance between the top surface of direct burial cable, conduit, or other raceways and the finished surface.

Wiring Method	Minimum Burial (Inches)
Direct burial cables	24
Rigid metal conduit or Intermediate metal conduit	6*
Eectrical metallic tubing	6*
Rigid nonmetallic conduit (Approved for direct burial Without concrete encasement)	18*

*For most locations, for complete details, refer to *NEC®* Table 300.5 for exceptions such as highways, dwellings, airports, driveways and parking lots.
Source: Data from NFPA 70®, *National Electrical Code®*, 2023 edition, NFPA. Quincy. MA., 2022, Table 300.5.

VOLUME REQUIRED PER CONDUCTOR

Size of Conductor	Free Space within Box for Each Conductor
No. 18	1.5 cubic inches
No. 16	1.75 cubic inches
No. 14	2 cubic inches
No. 12	2.25 cubic inches
No. 10	2.5 cubic inches
No. 8	3 cubic inches
No. 6	5 cubic inches

For complete details see *NEC* 314.16(B)
Source: NFPA 70®, *National Electrical Code®*. 2023 edition, NFPA. Quincy. MA. 2022, Table 314 16(B as modified.

SUPPORTING CONDUCTORS IN VERTICAL RACEWAYS — SPACINGS FOR CONDUCTOR SUPPORTS

Conductor Size	Support of Conductors in Vertical Raceways	Conductors Aluminum or Copper-Clad Aluminium	Copper
18 AWG through 8 AWG	Not greater than	100 feet	100 feet
6 AWG through 1/0 AWG	Not greater than	200 feet	100 feet
2/0 AWG through 4/0 AWG	Not greater than	180 feet	80 feet
Over 4/0 AWG through 350 kcmil	Not greater than	135 feet	60 feet
Over 350 kcmil through 500 kcmil	Not greater than	120 feet	50 feet
Over 500 kcmil through 750 kcmil	Not greater than	95 feet	40 feet
Over 750 kcmil	Not greater than	85 feet	35 feet

For SI units: 1 foot=0.3048 meter
Source: NFPA 70®, *National Electrical Code®*. 2023 edition, NFPA. Quincy. MA. 2022, Table 300 19(A). as modified.

MINIMUM DEPTH OF CLEAR WORKING SPACE AT ELECTRICAL EQUIPMENT

Nominal Voltage to Ground	Conditions		
	1	2	3
	Minimum Clear Distance (ft)		
0-150 V	3	3	3
151-600 V	3	3½	4
601-2500 V	3	4	5
2501-9000 V	4	5	6
9001-25000 V	5	6	9
25001 V-75 KV	6	8	10
Above 75 KV	8	10	12

Notes:

1. For SI units, 1 foot = 0.3048 meter.

2. Where the conditions are as follows:
   Condition 1—Exposed live parts on one side of the working space and no live or grounded parts on the other side of the working space, or exposed live parts on both sides of the working space that are effectively guarded by insulating materials.
   Condition 2—Exposed live parts on one side of the working space and grounded parts on the other side of the working space. Concrete, brick, or tile walls shall be considered as grounded.
   Condition 3—Exposed live parts on both sides of the working space.

See *Ugly's* pages 139–140 for metric conversions. For electrical rooms, where the equipment is rated 800 amps or more, *NEC* 110.26(C)(3) requires personnel doors to open at least 90 degrees in the egress direction and be equipped with listed panic hardware or listed fire exit hardware. Entrance to rooms shall meet the requirements of 110.27(C) and shall be clearly marked with warning signs forbidding unqualified persons to enter.

MINIMUM CLEARANCE OF LIVE PARTS, OVER 1000 VOLTS AC, 1500 VOLTS DC, NOMINAL

Nominal Voltage Rating KV	Impulse Withstand (BIL) (KV)		Minimum Clearance of Live Parts, Inches*			
			Phase-to-Phase		Phase-to-Ground	
	Indoors	Outdoors	Indoors	Outdoors	Indoors	Outdoors
2.4–4.16	60	95	4.5	7	3.0	6
7.2	75	95	5.5	7	4.0	6
13.8	95	110	7.5	12	5.0	7
14.4	110	110	9.0	12	6.5	7
23	125	150	10.5	15	7.5	10
34.5	150	150	12.5	15	9.5	10
	200	200	18.0	18	13.0	13
46		200		18		13
		250		21		17
69		250		21		17
		350		31		25
115		550		53		42
138		550		53		42
		650		63		50
161		650		63		50
		750		72		58
230		750		72		58
		900		89		71
		1050		105		83

* For SI units: 1 inch = 25.4 millimeters.

The values given are the minimum clearance for rigid parts and bare conductors under favorable service conditions. They shall be increased for conductor movement or under unfavorable service conditions, or wherever space limitations permit. The selection of the associated impulse withstand voltage for a particular system voltage is determined by the characteristics of the overvoltage (surge) protective equipment.

See *Ugly's* pages 139–140 for metric conversions.

Source: NFPA 70®, *National Electrical Code*®, 2023 edition, NFPA, Quincy, MA, 2022, Table 495.24, as modified.

MINIMUM SIZE EQUIPMENT GROUNDING CONDUCTORS FOR GROUNDING RACEWAY AND EQUIPMENT

Rating or Setting of Automatic Overcurrent Device in Circuit Ahead of Equipment, Conduit, Etc., Not Exceeding (Amperes)	Size (AWG of kcmil)	
	Copper	Aluminium or Copper-Clad Aluminium*
15	14	12
20	12	10
60	10	8
100	8	6
200	6	4
300	4	2
400	3	1
500	2	1/10
600	1	2/0
800	1/10	3/0
1000	2/0	4/0
1200	3/0	250 kcmil
1600	4/0	350 kcmil
2000	250 kcmil	400 kcmil
2500	350 kcmil	600 kcmil
3000	400 kcmil	600 kcmil
4000	500 kcmil	750 kcmil
5000	700 kcmil	1200 kcmil
6000	800 kcmil	1200 kcmil

Note: Where necessary to comply with *NEC* 250.4(A)(5) or 250.4(B)(4), the equipment grounding conductor shall be sized larger than given in this table.

*See installation restriction in *NEC* 250.120.

Source: NFPA 70®, *National Electrical Code®. 2023* edition, NFPA. Quincy, MA 2022. Table 250.122, as modified.

GROUNDING ELECTRODE CONDUCTOR FOR ALTERNATING-CURRENT SYSTEMS

Size of Largest Ungrounded conductor Equivalent Area for Parallel Conductors (AWG/kcmil)		Size of Grounding Electrode Conductor (AWG/kcmil)	
Copper	Aluminum or Copper-Clad Aluminum	Copper	Aluminum or Copper-Clad Aluminum
2 or Smaller	1/0 or Smaller	8	6
1 or 1/0	2/0 or 3/0	6	4
2/0 or 3/0	4/0 or 250	4	2
Over 3/0 through 350	Over 250 through 500	2	1/0
Over 350 through 600	Over 500 through 900	1/0	3/0
Over 600 through 1100	Over 900 through 1750	2/0	4/0
Over 1100	Over 1750	3/0	250

Notes:

1. If multiple sets of service-entrance conductors connect directly to a service drop, set of overhead service conductors, set of underground service conductors, or service lateral, the equivalent size of the largest service-entrance conductor shall be determined by the largest sum of the areas of the corresponding conductors of each set.

2. If there are no service-entrance conductors, the grounding electrode conductor size shall be determined by the equivalent size of the largest service-entrance conductor required for the load to be served.

3. See installation restrictions in *NEC* 250.64.

Source: NFPA 70®, *National Electrical Code*®, 2023 edition, NFPA, Quincy, MA, 2022, Table 250.66, as modified.

GROUNDED CONDUCTOR, MAIN BONDING JUMPER, SYSTEM BONDING JUMPER, AND SUPPLY-SIDE BONDING JUMPER FOR ALTERNATING-CURRENT SYSTEMS

Size of Largest Ungrounded Conductor or Equivalent Area for Parallel Conductors (AWG/kcmil)		Size of Grounded Conductor or Bonding Jumper* (AWG/kcmil)	
Copper	Aluminum or Copper-Clad Aluminum	Copper	Aluminum or Copper-Clad Aluminum
2 or Smaller	1/0 or Smaller	8	6
1 or 1/0	2/0 or 3/0	6	4
2/0 or 3/0	4/0 or 250	4	2
Over 3/0 through 350	Over 250 through 500	2	1/0
Over 350 through 600	Over 500 through 900	1/0	3/0
Over 600 through 1100	Over 900 through 1750	2/0	4/0
Over 1100	Over 1750	See Notes 1 and 2.	

Notes:
1. If the circular mil area of ungrounded supply conductors that are connected in parallel is larger than 1100 kcmil copper or 1750 kcmil aluminum, the grounded conductor or bonding jumper shall have an area not less than 12½% of the area of the largest ungrounded supply conductor or equivalent area for parallel supply conductors. The grounded conductor or bonding jumper shall not be required to be larger than the largest ungrounded supply conductor or set of ungrounded conductors.
2. If the circular mil area of ungrounded supply conductors that are connected in parallel is larger than 1100 kcmil copper or 1750 kcmil aluminum and if the ungrounded supply conductors and the bonding jumper are of different materials (copper, aluminum, or copper-clad aluminum), the minimum size of the grounded conductor or bonding jumper shall be based on the assumed use of ungrounded supply conductors of the same material as the grounded conductor or bonding jumper that has an ampacity equivalent to that of the installed ungrounded supply conductors.
3. If there are no service-entrance conductors, the supply conductor size shall be determined by the equivalent size of the largest service-entrance conductor required for the load to be served.

Source: NFPA 70®, *National Electrical Code®*, 2023 edition, NFPA, Quincy, MA, 2022, Table 250.102(C)(1), as modified.

GENERAL LIGHTING LOADS BY NON-DWELLING OCCUPANCY

Type of Occupancy	Volt-Amperes/ Square Foot	Type of Occupancy	Volt-Amperes/ Square Foot
Automotive facility	1.5	Museum	1.6
Convention center	1.4	Office[4]	1.3
Courthouse	1.4	Parking garage[5]	0.3
Dormitory	1.5	Penitentiary	1.2
Exercise center	1.4	Performing arts theater	1.5
Fire station	1.3	Police station	1.3
Gymnasium[1]	1.7	Post office	1.6
Health care clinic	1.6	Religious facility	2.2
Hospital	1.6	Restaurant[6]	1.5
Hotels & motels, including		Retails[7, 8]	1.9
apartment without provisions		School/university	1.5
for cooking by tenants[2]	1.7	Sports arena	1.5
Library	1.5	Town hall	1.4
Manufacturing facility[3]	2.2	Transportation	1.2
Motion picture theater	1.6	Warehouse	1.2
		Workshop	1.7

See *NEC* 220.41 for dwelling units.
See *NEC* 220.14(J) for receptacle outlets in office buildings.

Note: The 125% multiplier for a continuous load as specified in 210. 20(A) is included, therefore no additional multiplier shall be required when using the unit loads in this table for calculating the minimum lighting load for a specified occupancy.
[1]Armories and auditoriums are considered gymnasium-type occupancies.
[2]Lodge rooms are similar to hotels and motels.
[3]Industrial commercial loft buildings are considered manufacturing-type occupancies.
[4]Banks are office-type occupancies.
[5]Commercial (storage) garages are considered parking garage occupancies.
[6]Clubs are considered restaurant occupancies.
[7]Barber shops and beauty parlors are considered retail occupancies.
[8]Stores are considered retail occupancies.

Source: NFPA 70®, *National Electrical Code*®, 2023 edition, NFPA, Quincy, MA, 2022, Table 220.42(A), as modified.

LIGHTING LOAD DEMAND FACTORS

Type of Occupancy	Portion of Lighting Load to Which Demand Factor Applies (Volt-Amperes	Demand Factor (%)
Dwelling units	First 3000	100
	From 3001 to 120000 at	35
	Remainder over 120000 at	25
Hotels and motels, including apartment houses without provision for cooking by tenants*	First 20000 or less at	60
	From 20001 to 100000 at	50
	Remainder over 100000 at	35
Warehouses (storage)	First 12500 or less at	100
	Remainder over 12500 at	50
All others	Total volt-amperes	100

* The demand factors of this table shall not apply to the calculated load of feeders or services supplying areas in hotels and motels where the entire lighting is likely to be used at one time, as in ballrooms or dining rooms.

Source: NFPA 70®, *National Electrical Code*®, 2023 edition, NFPA, Quincy, MA, 2022. Table 220.45, as modified.

 # DEMAND FACTORS FOR RECEPTACLE LOADS—OTHER THAN DWELLING UNITS

Portion of Receptacle Load to Which Demand Factor Applies (Volt-Amperes	Demand Factor (%)
First 10 kVA or less at	100
Remainder over 10 kVA at	50

Source: NFPA 70®, *National Electrical Code*®, 2023 edition, NFPA, Quincy, MA, 2022, Table 220.42, as modified.

DEMAND FACTORS FOR HOUSEHOLD ELECTRIC CLOTHES DRYERS

Number of Dryers	Demand Factor
1-4	100%
5	85%
6	75%
7	65%
8	60%
9	55%
10	50%
11	47%
12-23	47% minus 1% for each dryer exceeding 11
24-42	35% minus 0.5% for each dryer exceeding 23
43 and over	25%

Source: NFPA 70®, *National Electrical Code*®, 2023 edition, NFPA, Quincy, MA, 2022, Table 220.54, as modified.

DEMAND FACTORS FOR KITCHEN EQUIPMENT—OTHER THAN DWELLING UNIT(S)

Number of Units of Equipment	Demand Factor (%)
1	100
2	100
3	90
4	80
5	70
6 and over	65

Note: In no case shall the feeder or service calculated load be less than the sum of the largest two kitchen equipment loads.

Source: NFPA 70®, *National Electrical Code*®, 2023 edition, NFPA, Quincy, MA, 2022, Table 220.56, as modified.

DEMAND LOADS FOR HOUSEHOLD ELECTRIC RANGES, WALL-MOUNTED OVENS, COUNTER-MOUNTED COOKING UNITS, AND OTHER HOUSEHOLD COOKING APPLIANCES OVER 1¾ KW RATING

(Column C to be used in all cases except as otherwise permitted in Note 3)

Number of Appliances	Demand Factor (%) (See Notes)		Column C Maximum Demand (kW) (See Notes) (Not over 12 kW Rating)
	Column A (Less than 3 ½ kW Rating)	Column B (3 ½ kW Through 8¾ kW Rating)	
1	80	80	8
2	75	65	11
3	70	55	14
4	66	50	17
5	62	45	20
6	59	43	21
7	56	40	22
8	53	36	23
9	51	35	24
10	49	34	25
11	47	32	26
12	45	32	27
13	43	32	28
14	41	32	29
15	40	32	30
16	39	28	31
17	38	28	32
18	37	28	33
19	36	28	34
20	35	28	35
21	34	26	36
22	33	26	37
23	32	26	38
24	31	26	39
25	30	26	40
26-30	30	24 ⟩	15 kW + 1 kW for each range
31-40	30	22	
41-50	30	20 ⟩	25 kW + 3/4 kW for each range
51-60	30	18	
61 and over	30	16	

(continued on next page)

115

DEMAND LOADS FOR HOUSEHOLD ELECTRIC RANGES, WALL-MOUNTED OVENS, COUNTER-MOUNTED COOKING UNITS, AND OTHER HOUSEHOLD COOKING APPLIANCES OVER 1¾ KW RATING

Notes:

1. Over 12 kW through 27 kW ranges all of same rating. For ranges individually rated more than 12 kW but not more than 27 kW, the maximum demand in Column C shall be increased by 5% for each additional kilowatt of rating or major fraction thereof by which the rating of individual ranges exceeds 12 kW.

2. Over 8¾ kW through 27 kW ranges of unequal ratings. For ranges individually rated more than 8¾ kW and of different ratings, but none exceeding 27 kW, an average value of rating shall be computed by adding together the ratings of all ranges to obtain the total connected load (using 12 kW for any range rated less than 12 kW) and dividing the total number of ranges. Then the maximum demand in Column C shall be increased by 5% for each kilowatt or major fraction thereof by which this average value exceeds 12 kW.

3. Over 1¾ kW through 8¾ kW. In lieu of the method provided in Column C, adding the nameplate ratings of all household cooking appliances rated more than 1¾ kW but not more than 8¾ kW and multiplying the sum by the demand factors specified in Column A or B for the given number of appliances shall be permitted. Where the rating of cooking appliances falls under both Column A and Column B, the demand factors for each column shall be applied to the appliances for that column and the results are added together.

4. Calculating the branch-circuit load for one range in accordance with Table 220.55 shall be permitted.

5. The branch-circuit load for one wall-mounted oven or one counter-mounted cooking unit shall be the nameplate rating of the appliance.

6. The branch-circuit load for a counter-mounted cooking unit and not more than two wall-mounted ovens, all supplied from a single branch circuit and located in the same room, shall be computed by adding the nameplate rating of the individual appliances and treating this total as equivalent to one range.

7. This table also applies to household cooking appliances rated over 1¾ kW and used in instructional programs.

NFPA 70®, National Electrical Code®, 2023 edition, NFPA, Quincy, MA, 2022. Table 220.55, as modified.

CALCULATING COST OF OPERATING AN ELECTRICAL APPLIANCE

What is the monthly cost of operating a 240-volt, 5-kilowatt (kW) central electric heater that operates 12 hours per day when the cost is 15 cents per kilowatt-hour (kWhr)?

$$\text{Cost} = \text{Watts} \times \text{Hours Used} \times \text{Rate per kWhr/1000}$$

5 kW = 5000 Watts

Hours = 12 Hours $\times$ 30 Days = 360 Hours per Month

= 5000 $\times$ 360 $\times$ 0.15/1000

= 270000/1000 = **$270 Monthly Cost**

The above example is for a resistive load. Air-conditioning loads are primarily inductive loads. However, if ampere and voltage values are known, this method will give an approximate cost. Kilowatt-hour rates vary for different power companies, and for residential use, graduated-rate scales are usually used (the more power used, the lower the rate). Commercial and industrial rates are generally based on kilowatt usage, maximum demand and power factor. Other costs are often added such as fuel cost adjustments.

CHANGING INCANDESCENT LAMP TO ENERGY-SAVING LAMP

A 100-watt incandescent lamp is to be replaced with a 15-watt, energy-saving lamp that has the same light output (lumens). If the cost per kilowatt-hour (kWhr) is 15 cents, how many hours would the new lamp need to operate to pay for itself?

Lamp cost is 4 dollars. Energy saved is 85 watts.

$$\text{Hours} = \text{Lamp Cost} \times 1000/\text{Watts Saved} \times \text{kWhr}$$

(4 $\times$ 1000)/(85 $\times$ 0.15) = 4000/12.75 = 313.73 hours

The energy-saving lamp will pay for itself with 313.73 hours of operation.

The comparative operating cost of these two lamps based on 313.73 hours is found by:

$$\text{Cost} = \text{Watts} \times \text{Hours Used} \times \text{Rate per kWhr/1000}$$

100-watt incandescent lamp = $4.71 for 313.73 hours of operation

15-watt energy saving lamp = $0.71 for 313.73 hours of operation

PARTIAL 2023 NATIONAL ELECTRICAL CODE SUMMARY

The 2023 edition of the National Electrical Code (NEC) contains 13 new articles but seven existing articles were deleted. With most of the articles that were deleted, requirements were moved into new articles. Article 510 was deleted but the necessary text from that article was added to Articles 511 through 516. Before Article 712 was deleted, the Code-Making Panel reviewed the language in each section and compared those with requirements located in Article 705 as well as other related articles, including those in Chapters 1 through 4. Article 720, which was added to the Code in 1920, has been deleted because the requirements no longer apply.

Article 235—Branch Circuits, Feeders, and Services Over 1000 Volts ac, 1500 Volts dc, Nominal (New)

Article 235 contains general requirements for branch circuits, feeders, and services over 1000 volts ac or 1500 volts dc, nominal.

Article 245—Overcurrent Protection for Systems Rated Over 1000 Volts ac, 1500 Volts dc (New)

Article 245 covers overcurrent protection requirements for systems over 1000 volts ac, 1500 volts dc, nominal.

Article 305—General Requirements for Wiring Methods and Materials for Systems Rated Over 1000 Volts ac, 1500 Volts de, Nominal (New)

Article 305 covers wiring methods and materials for systems rated over 1000 volts ac, 1500 volts dc, nominal.

Article 315—Medium Voltage Conductors, Cables, Cable Joints, and Cable Terminations (New)

Article 315 covers the use, installation, construction specifications, and ampacities for Type MV medium voltage conductors, cable, cable joints, and cable terminations. Article 315 includes voltages from 2001 volts to 35,000 volts ac, nominal and 2001 volts to 2500 volts dc, nominal. Article 311, which was new in the 2020 edition, was deleted from the 2023 *NEC* and the requirements were moved into new Article 315.

PARTIAL 2023 NATIONAL ELECTRICAL CODE SUMMARY

Article 335—Instrumentation Tray Cable: Type ITC (New)

Article 335 covers the use, installation, and construction specifications of instrumentation tray cable (Type ITC) for application to instrumentation and control circuits operating at 150 volts or less and 5 amperes or less. Former Article 727 was deleted from the 2023 *NEC* and the requirements were moved into the new Article 335.

Article 369—Insulated Bus Pipe (IBP)/Tubular Covered Conductors (TCC) Systems (New)

Article 369 covers the use, installation, and construction specifications for insulated bus pipe (IBP) systems.

Article 371—Flexible Buss Systems (New)

Article 371 covers the use and installation requirements of flexible bus systems and associated fittings.

Article 395—Outdoor Overhead Conductors over 1000 Volts (New)

Article 395 covers the use and installation for outdoor overhead conductors over 1000 volts, nominal. Former Article 399 was deleted from the 2023 *NEC* and the requirements were moved into the new Article 395.

Article 495—Equipment Over 1000 Volts ac, 1500 Volts de, Nominal (New)

Article 495 covers the general requirements for equipment operating at more than 1000 volts ac, 1500 volts dc, nominal. Former Article 490 was deleted from the 2023 *NEC* and the requirements were moved into the new Article 495.

Article 512—Cannabis Oil Equipment and Cannabis Oil Systems Using Flammable Materials (New)

Article 512 covers cannabis oil preparatory equipment, extraction equipment, booths, post-processing equipment, and systems using flammable materials (flammable gas, flammable liquid–produced vapor, combustible liquid–produced vapor) in commercial and industrial facilities.

PARTIAL 2023 NATIONAL ELECTRICAL CODE SUMMARY

Article 722—Cables for Power-Limited Circuits and Fault-Managed Power Circuits (New)

Article 722 covers the general requirements for the installation of single- and multiple-conductor cables used in Class 2 and Class 3 power-limited circuits, power-limited fire alarm (PLFA) circuits, and Class 4 fault-managed power circuits. Common cabling requirements found in Articles 725 and 760 of the 2020 NEC have been relocated into new Article 722 within the 2023 NEC.

Article 724—Class 1 Power-Limited Circuits and Class 1 Power-Limited Remote-Control and Signaling Circuits (New)

Article 724 covers Class 1 circuits, including power-limited Class 1 remote-control and signaling circuits, that are not an integral part of a device or utilization equipment.

Article 726—Class 4 Fault-Managed Power Systems (New)

Article 726 covers the installation of wiring systems and equipment, including utilization equipment, of Class 4 fault-managed power (FMP) systems.

Definitions [Article 100]

Part 1. General, Part II. Over 1000 Volts, Nominal, and Part III Hazardous (Classified) Locations have been removed from Article 100. In previous editions, if a word or term was only used in one article, the definition would be located in that specific article. Now, all definitions are in Article 100 and are arranged in alphabetical order. An article number in parentheses following the definition indicates that the definition only applies to that one article.

Arc-Flash Hazard Warning [110.16]

Besides the arc-flash label being required on service equipment, it is also required on feeder supplied equipment. The amperage rating was also lowered from 1200 amps to 1000 amps. The arc-flash hazard warning label is still not required for dwelling units. Besides the arc-flash label being required on service equipment, it is also required on feeder supplied equipment. The amperage rating was also lowered from 1200 amps or more to 1000 amps or more. The arc-flash hazard warning label is still not required for dwelling units. The list of items required on the label has been deleted and now this requirement says the arc flash label shall be in accordance with applicable industry practice and include the date the label was applied. Revised Informational Note No. 2 lists NFPA 70E *Standard for Electrical Safety in the Workplace* as being a resource for acceptable industry practices for equipment labeling.

GFCI Protection for Dwelling Unit Kitchen Receptacles [210.8(A)]

Previously, the wording in 210.8(A)(6) stated receptacles installed to serve countertop surfaces in kitchens had to be GFCI protected. Now in the 2023 edition, all 125-volt through 250-volt receptacles installed in kitchens are required to be GFCI protected. Before the 2023 edition, a 120-volt receptacle installed on a wall more than 6 feet from a sink and not installed to serve countertop surfaces did not have to be GFCI protected. Now, because the receptacle is located in the kitchen, it is required to be GFCI protected. Because this section also applies to 240-volt receptacles, a receptacle supplying power to an electric range is also required to be GFCI protected.

Arc-Fault Circuit-Interrupter Protection [210.12]

Within 210.12(D), the required locations for arc-fault circuit-interrupter protection (AFCI) have been expanded to include areas designed for use exclusively as sleeping quarters in fire stations, police stations, ambulance stations, rescue stations, ranger stations, and similar locations. AFCI protection has also been expanded to include 120-volt, single-phase, 10-ampere circuits the aforementioned areas as well as in dwelling units per 210.12(B) and dormitory units per 210.12(C).

10-Ampere Branch Circuits [210.23(A)]

Branch circuit ratings now include 10 amperes. As specified in 210.23(A), a 10-ampere branch circuit shall comply with the requirements of 210.23(A)(1) and (A)(2). Permitted loads include lighting outlets, dwelling unit exhaust fans on bathroom or laundry room lighting circuits, and a gas fireplace unit supplied by an individual branch circuit. Loads not permitted for 10-ampere branch circuits include receptacle outlets; fixed appliances, except as permitted for individual branch circuits; garage door openers; and laundry equipment. Table 240.6(A) now includes 10 amperes as a standard ampere rating for fuses and inverse time circuit breakers.

Feeder Surge Protection [215.18]

A surge-protection device (SPD) having a nominal discharge current rating (In) of not less than 10kA, shall now be installed in or adjacent to the distribution equipment containing branch circuit overcurrent protective device(s) that supply the location specified in 215.18(A). In accordance with 215.18(A), an SPD shall be installed on feeders in dwelling units, dormitory units, guest rooms and guest suites of hotels and motels, and areas of nursing homes and limited-care facilities used exclusively as patient sleeping rooms. Where the distribution equipment supplied by the feeder is replaced, all of the requirements in 215.18 apply.

Floor Area Calculations [220.5(C)]

When calculating the floor area to be used in branch circuit and feeder load calculations, we are and have been instructed to use the outside dimensions of the building. Prior to the 2023 edition, open porches, garages, or unused or unfinished spaces not adaptable for future use in dwelling units did not have to be included in floor area calculations. Besides this section being relocated to 220.5(C) from 220.11, this section was also reworded for clarity. The calculated floor area for dwelling units shall not include open porches or unfinished areas not adaptable for future use as a habitable room or occupiable space.

Electric Vehicle Supply Equipment (EVSE) Load [220.57]

A new requirement has been added for sizing loads for electric vehicle supply equipment. The EVSE load shall be calculated at either 7200 watts (volt-amperes) or the nameplate rating of the equipment, whichever is larger. Article 625 covers the electrical conductors and equipment connecting an electric vehicle to premises wiring for the purposes of charging, power export, or bidirectional current flow.

Tamper-Resistant Receptacles [406.12(1)]

Article 406 expanded the requirement for tamper-resistant receptacles. All 15- and 20-ampere, 125- and 250-volt nonlocking-type receptacles in the locations specified in 406.12(1) though (10) shall be listed tamper-resistant receptacles. The first group of locations now includes all dwelling units, boathouses, mobile homes and manufactured homes, including their attached and detached garages, accessory buildings, and common areas.

Nonmetallic Cable in Agricultural Buildings [547.26]

The first sentence in 547.26, which states all electrical wiring and equipment subject to physical damage shall be protected, did not change. A second sentence was added, which states nonmetallic cables shall not be permitted to be concealed within walls and above ceilings of buildings (i.e., offices, lunchrooms, ancillary areas, etc.) or portions thereof, which are contiguous with or physically adjoined to livestock confinement areas. The new Informational Note explains why this requirement is necessary. Rodents and other pests are common around such installations and will damage nonmetallic cable by chewing the cable jacket and conductor insulation concealed within walls and ceilings of livestock containment areas of agricultural buildings.

FIELD TERMS VERSUS *NEC* TERMS

- BX   Armored cable (*NEC* 320)
- Romex   Nonmetallic sheathed cable (*NEC* 334)
- Green field   Flexible metal conduit (*NEC* 348)
- Thin wall   Electrical metallic tubing (*NEC* 358)
- Smurf tube   Electrical nonmetallic tubing (*NEC* 362)
- 1900 box   4-inch square box (*NFC* 314)
- 333 box   Device box (*NEC* 314)
- EYS   Explosion proof seal off (*NEC* 500)
- Neutral**   Grounded conductor (*NEC* 200)**
- Ground wire   Equipment grounding conductor (*NEC* 250.118)
- Ground wire   Grounding electrode conductor (*NEC*250.66)
- Hot, live   Ungrounded conductor (*NEC* 100)

** Some systems do not have a neutral, and the grounded conductor may be a phase conductor. (See *NEC* Article 100 definition of *neutral conductor.*)

ELECTRICAL SYMBOLS

Wall	Ceiling	
—○	○	Outlet
—Ⓓ	Ⓓ	Drop Card
—Ⓕ	Ⓕ	Fan Outlet
—Ⓙ	Ⓙ	Junction Box
—Ⓛ	Ⓛ	Lampholder
—Ⓛ PS	Ⓛ PS	Lampholder With Pull Switch
—Ⓢ	Ⓢ	Pull Switch
—Ⓒ	Ⓒ	Vapor Discharge Lamp Outlet
—Ⓧ	Ⓧ	Exit Outlet
—Ⓒ	Ⓒ	Clock Outlet
—Ⓑ	Ⓑ	Blanked Outlet

⊜ Duplex Convenience Outlet

⊜ 1.3 Single, Triplex, etc.

⊜ Range Outlet

⊝ S Duplex Receptacle and Switch

▲ Special Purpose Outlet

● Floor Outlet

Switch Outlets

S	Single-Pole Switch
S_2	Double-Pole Switch
S_3	Three-Way Switch
S_4	Four-Way Switch
S_D	Door Switch
S_E	Electrolier Switch
S_P	Switch and Pilot Lamp
S_K	Key-Operated Switch
S_{CB}	Circuit Breaker
S_{WCB}	Weather-Proof Circuit Breaker
S_{MC}	Momentary-Contact Switch
S_{RC}	Remote-Control Switch
S_{WP}	Weather-Proof Switch
S_F	Fused Switch
S_{WPF}	Weather-Proof Fused Switch
▬	Lighting Switch
▨	Power Panel

ELECTRICAL SYMBOLS

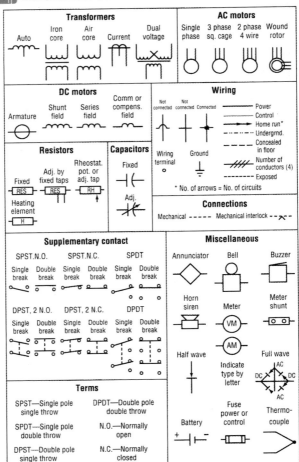

Transformers

Auto	Iron core	Air core	Current	Dual voltage

AC motors

Single phase | 3 phase sq. cage | 2 phase 4 wire | Wound rotor

DC motors

Armature	Shunt field	Series field	Comm or compens. field

Wiring

Not connected | Not connected | Connected

——— Power
----- Control
——➤ Home run*
-··-··- Undergrnd.
------- Concealed in floor
Wiring terminal ○ Ground
//// Number of conductors (4)
------- Exposed
* No. of arrows = No. of circuits

Resistors

Fixed	Adj. by fixed taps	Rheostat. pot. or adj. tap
RES	RES	RH

Heating element
H

Capacitors

Fixed

Adj.

Connections

Mechanical ----- Mechanical interlock --✕-

Supplementary contact

SPST.N.O.
Single break | Double break

SPST.N.C.
Single break | Double break

SPDT
Single break | Double break

DPST, 2 N.O.
Single break | Double break

DPST, 2 N.C.
Single break | Double break

DPDT
Single break | Double break

Miscellaneous

Annunciator | Bell | Buzzer

Horn siren | Meter | Meter shunt
VM

AM

Half wave

Indicate type by letter

Full wave
AC DC DC AC

Battery

Fuse power or control

Thermo-couple

Terms

SPST—Single pole single throw

SPDT—Single pole double throw

DPST—Double pole single throw

DPDT—Double pole double throw

N.O.—Normally open

N.C.—Normally closed

ELECTRICAL SYMBOLS

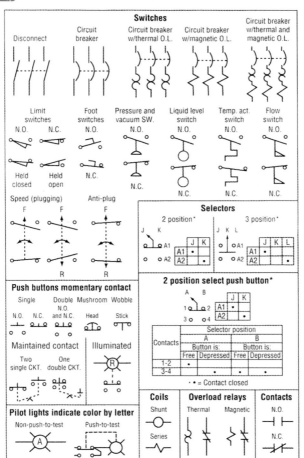

Switches

Disconnect | Circuit breaker | Circuit breaker w/thermal O.L. | Circuit breaker w/magnetic O.L. | Circuit breaker w/thermal and magnetic O.L.

Limit switches (N.O., N.C.) | Foot switches (N.O., N.C.) | Pressure and vacuum SW. (N.O., N.C.) | Liquid level switch (N.O., N.C.) | Temp. act. switch (N.O., N.C.) | Flow switch (N.O., N.C.)

Held closed | Held open

Speed (plugging) F, F, F ... R, R | Anti-plug F

Selectors

2 position*

	J	K
A1	•	
A2		•

3 position*

	J	K	L
A1	•		
A2			•

Push buttons momentary contact

Single (N.O., N.C.) | Double N.O. N.C. | Mushroom and N.C. Head | Wobble Stick

Maintained contact

Two single CKT. | One double CKT.

Illuminated

2 position select push button*

	J	K
A1	•	
A2		•

Contacts	Selector position			
	A		B	
	Button is:		Button is:	
	Free	Depressed	Free	Depressed
1-2	•			
3-4		•	•	•

• = Contact closed

Pilot lights indicate color by letter

Non-push-to-test | Push-to-test

Coils	Overload relays		Contacts
Shunt	Thermal	Magnetic	N.O.
Series			N.C.

Note: N.O. = normally open; N.C. = normally closed

WIRING DIAGRAMS FOR NEMA CONFIGURATIONS

2 Pole, 2 Wire
Nongrounding
125V

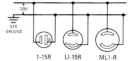

1-15R   LI-15R   ML1-R

2 Pole, 2 Wire
Nongrounding
250V

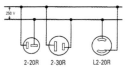

2-20R   2-30R   L2-20R

2 Pole, 3 Wire
Grounding
125V

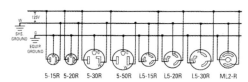

5-15R 5-20R 5-30R 5-50R L5-15R L5-20R L5-30R ML2-R

2 Pole, 3 Wire
Grounding
250V

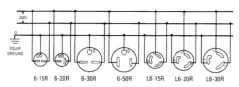

6-15R 6-20R 6-30R 6-50R L6-15R L6-20R L6-30R

2 Pole, 3 Wire
Grounding
277V AC

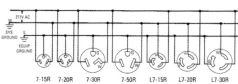

7-15R 7-20R 7-30R 7-50R L7-15R L7-20R L7-30R

WIRING DIAGRAMS FOR NEMA CONFIGURATIONS

2 Pole, 3 Wire Grounding 480V AC

L8-20R   L8-30R

3 Pole, 3 Wire Nongrounding 125/250V

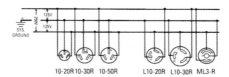

10-20R 10-30R 10-50R   L10-20R L10-30R ML3-R

3 Pole, 3 Wire Nongrounding 3ø 250V

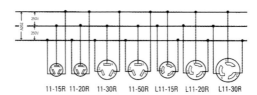

11-15R 11-20R 11-30R   11-50R   L11-15R L11-20R L11-30R

3 Pole, 4 Wire Grounding 125/250V

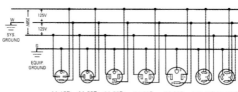

14-15R 14 20R 14-30R   14-50R   14-60R   L14-20R L14-30R

3 Pole, 4 Wire
Grounding
3ø 250V

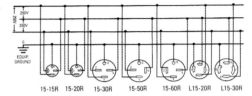

15-15R   15-20R   15-30R   15-50R   15-60R   L15-20R   L15-30R

3 Pole, 4 Wire
Grounding
3ø 480V

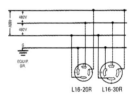

L16-20R   L16-30R

3 Pole, 4 Wire
Grounding
3ø 600V

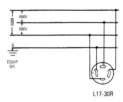

L17-30R

WIRING DIAGRAMS FOR NEMA CONFIGURATIONS

**4 Pole, 4 Wire
Nongrounding
3ø 120/208V**

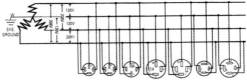

18-15R 18-20R 18-30R 18-50R 18-60R L18-20R L18-30R

**4 Pole, 4 Wire
Nongrounding
3ø 277/480V**

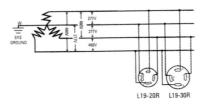

L19-20R L19-30R

**4 Pole, 4 Wire
Nongrounding
3ø 347/600V**

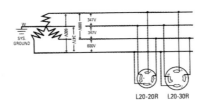

L20-20R L20-30R

4 Pole, 5 Wire Grounding 3ø 120/208V

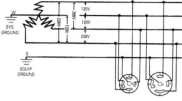

L21-20R   L21-30R

4 Pole, 5 Wire Grounding 3ø 277/480V

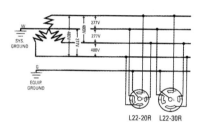

L22-20R   L22-30R

4 Pole, 5 Wire Grounding 3ø 347/600V

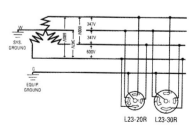

L23-20R   L23-30R

 **NEMA ENCLOSURE TYPES**

The specific enclosure types, their applications, and the environmental conditions they are designed to provide a degree of protection against are as follows:

Type 1—Enclosures constructed for indoor use to provide a degree of protection to personnel against access to hazardous parts and to provide a degree of protection of the equipment inside the enclosure against ingress of solid foreign objects (falling dirt).

Type 2—Enclosures constructed for indoor use to provide a degree of protection to personnel against access to hazardous parts; to provide a degree of protection of the equipment inside the enclosure against ingress of solid foreign objects (falling dirt); and to provide a degree of protection with respect to harmful effects on the equipment due to the ingress of water (dripping and light splashing).

Type 3—Enclosures constructed for either indoor or outdoor use to provide a degree of protection to personnel against access to hazardous parts; to provide a degree of protection of the equipment inside the enclosure against ingress of solid foreign objects (falling dirt and windblown dust); to provide a degree of protection with respect to harmful effects on the equipment due to the ingress of water (rain, sleet, snow); and that will be undamaged by the external formation of ice on the enclosure.

Type 3R—Enclosures constructed for either indoor or outdoor use to provide a degree of protection to personnel against access to hazardous parts; to provide a degree of protection of the equipment inside the enclosure against ingress of solid foreign objects (falling dirt); to provide a degree of protection with respect to harmful effects on the equipment due to the ingress of water (rain, sleet, snow); and that will be undamaged by the external formation of ice on the enclosure.

Type 3S—Enclosures constructed for either indoor or outdoor use to provide a degree of protection to personnel against access to hazardous parts; to provide a degree of protection of the equipment inside the enclosure against ingress of solid foreign objects (falling dirt and windblown

Reprinted by permission of the National Electrical manufacturers Association, Electrical Enclosure Types – Non Hazardous Location Environmental Rating Standards Comparison.

dust); to provide a degree of protection with respect to harmful effects on the equipment due to the ingress of water (rain, sleet, snow); and for which the external mechanism(s) remain(s) operable when ice laden.

Type 3X—Enclosures constructed for either indoor or outdoor use to provide a degree of protection to personnel against access to hazardous parts; to provide a degree of protection of the equipment inside the enclosure against ingress of solid foreign objects (falling dirt and windblown dust); to provide a degree of protection with respect to harmful effects on the equipment due to the ingress of water (rain, sleet, snow); that provides an increased level of protection against corrosion and that will be undamaged by the external formation of ice on the enclosure.

Type 3RX—Enclosures constructed for either indoor or outdoor use to provide a degree of protection to personnel against access to hazardous parts; to provide a degree of protection of the equipment inside the enclosure against ingress of solid foreign objects (falling dirt); to provide a degree of protection with respect to harmful effects on the equipment due to the ingress of water (rain, sleet, snow); that will be undamaged by the external formation of ice on the enclosure that provides an increased level of protection against corrosion.

Type 3SX—Enclosures constructed for either indoor or outdoor use to provide a degree of protection to personnel against access to hazardous parts; to provide a degree of protection of the equipment inside the enclosure against ingress of solid foreign objects (falling dirt and windblown dust); to provide a degree of protection with respect to harmful effects on the equipment due to the ingress of water (rain, sleet, snow); that provides an increased level of protection against corrosion; and for which the external mechanism(s) remain(s) operable when ice laden.

Type 4—Enclosures constructed for either indoor or outdoor use to provide a degree of protection to personnel against access to hazardous parts; to provide a degree of protection of the equipment inside the enclosure against ingress of solid foreign objects (falling dirt

Reprinted from *NEMA 250-2020* by permission of the National Electrical Manufacturers Association.

NEMA ENCLOSURE TYPES

and windblown dust); to provide a degree of protection with respect to harmful effects on the equipment due to the ingress of water (rain, sleet, snow, splashing water, and hose-directed water); and that will be undamaged by the external formation of ice on the enclosure.

Type 4X—Enclosures constructed for either indoor or outdoor use to provide a degree of protection to personnel against access to hazardous parts; to provide a degree of protection of the equipment inside the enclosure against ingress of solid foreign objects (falling dirt and windblown dust); to provide a degree of protection with respect to harmful effects on the equipment due to the ingress of water (rain, sleet, snow, splashing water, and hose-directed water); that provides an increased level of protection against corrosion; and that will be undamaged by the external formation of ice on the enclosure.

Type 5—Enclosures constructed for indoor use to provide a degree of protection to personnel against access to hazardous parts; to provide a degree of protection of the equipment inside the enclosure against ingress of solid foreign objects (falling dirt and settling airborne dust, lint, fibers, and flyings); and to provide a degree of protection with respect to harmful effects on the equipment due to the ingress of water (dripping and light splashing).

Type 6—Enclosures constructed for either indoor or outdoor use to provide a degree of protection to personnel against access to hazardous parts; to provide a degree of protection of the equipment inside the enclosure against ingress of solid foreign objects (falling dirt); to provide a degree of protection with respect to harmful effects on the equipment due to the ingress of water (hose-directed water and the entry of water during occasional temporary submersion at a limited depth); and that will be undamaged by the external formation of ice on the enclosure.

Type 6P—Enclosures constructed for either indoor or outdoor use to provide a degree of protection to personnel against access to hazardous parts; to provide a degree of protection of the equipment inside the

🔌 NEMA ENCLOSURE TYPES

enclosure against ingress of solid foreign objects (falling dirt); to provide a degree of protection with respect to harmful effects on the equipment due to the ingress of water (hose-directed water and the entry of water during prolonged submersion at a limited depth); that provides an increased level of protection against corrosion; and that will be undamaged by the external formation of ice on the enclosure.

Type 12—Enclosures constructed (without knockouts) for indoor use to provide a degree of protection to personnel against access to hazardous parts; to provide a degree of protection of the equipment inside the enclosure against ingress of solid foreign objects (falling dirt and circulating dust, lint, fibers, and flyings); to provide a degree of protection with respect to harmful effects on the equipment due to the ingress of water (dripping and light splashing); and to provide a degree of protection against light splashing and seepage of oil and non-corrosive coolants.

Type 12K—Enclosures constructed (with knockouts) for indoor use to provide a degree of protection to personnel against access to hazardous parts; to provide a degree of protection of the equipment inside the enclosure against ingress of solid foreign objects (falling dirt and circulating dust, lint, fibers, and flyings); to provide a degree of protection with respect to harmful effects on the equipment due to the ingress of water (dripping and light splashing); and to provide a degree of protection against light splashing and seepage of oil and non-corrosive coolants.

Type 13—Enclosures constructed for indoor use to provide a degree of protection to personnel against access to hazardous parts; to provide a degree of protection of the equipment inside the enclosure against ingress of solid foreign objects (falling dirt and circulating dust, lint, fibers, and flyings); to provide a degree of protection with respect to harmful effects on the equipment due to the ingress of water (dripping and light splashing); and to provide a degree of protection against the spraying, splashing, and seepage of oil and non-corrosive coolants.

Reprinted from *NEMA 250-2020* by permission of the National Electrical Manufacturers Association.

🔲 U.S. WEIGHTS AND MEASURES

Linear Measures

	1 Inch	= 2.540 Centimeters
12 Inches	= 1 Foot	= 3.048 Decimeters
3 Feet	= 1 Yard	= 9.144 Decimeters
5.5 Yards	= 1 Rod	= 5.029 Meters
40 Rods	= 1 Furlong	= 2.018 Hectometers
8 Furlongs	= 1 Mile	= 1.609 Kilometers

Mile Measurements

1 Statute Mile	= 5280 Feet
1 Scots Mile	= 5952 Feet
1 Irish Mile	= 6720 Feet
1 Russian Verst	= 3504 Feet
1 Italian Mile	= 4401 Feet
1 Spanish Mile	= 15084 Feet

Other Linear Measurements

1 Hand	=	4 Inches		1 Link	=	7.92 Inches
1 Span	=	9 Inches		1 Fathom	=	6 Feet
1 Chain	=	22 Yards		1 Furlong	=	10 Chains
				1 Cable	=	608 Feet

Square Measures

144	Square Inches	= 1	Square Foot
9	Square Feet	= 1	Square Yard
30¼	Square Yards	= 1	Square Rod
40	Rods	= 1	Rood
4	Roods	= 1	Acre
640	Acres	= 1	Square Mile
1	Square Mile	= 1	Section
36	Sections	= 1	Township

Cubic or Solid Measures

1	Cubic Foot	=	1728	Cubic Inches
1	Cubic Yard	=	27	Cubic Feet
1	Cubic Foot	=	7.48	Gallons
1	Gallon (Water)	=	8.34	Pounds
1	Gallon (U.S.)	=	231	Cubic Inches of Water
1	Gallon (Imperial)	=	277¼	Cubic Inches of Water

U.S. WEIGHTS AND MEASURES

Liquid Measurements

1 Pint	=	4 Gills
1 Quart	=	2 Pints
1 Gallon	=	4 Quarts
1 Firkin	=	9 Gallons (Ale or Beer)
1 Barrel	=	42 Gallons (Petroleum or Crude Oil)

Dry Measure

1 Quart	= 2 Pints
1 Peck	= 8 Quarts
1 Bushel	= 4 Pecks

Weight Measurement (Mass)

A. Avoirdupois Weight

1 Ounce	=	16 Drams
1 Pound	=	16 Ounces
1 Hundredweight	=	100 Pounds
1 Ton	=	2000 Pounds

B. Troy Weight

1 Carat	=	3.17 Grains
1 Pennyweight	=	20 Grains
1 Ounce	=	20 Pennyweights
1 Pound	=	12 Ounces
1 Long Hundred-Weight	=	112 Pounds
1 Long Ton	=	20 Long Hundredweights
	=	2240 Pounds

C. Apothecaries Weight

1 Scruple	= 20	Grains	=	1.296	Grams
1 Dram	= 3	Scruples	=	3.888	Grams
1 Ounce	= 8	Drams	=	31.1035	Grams
1 Pound	= 12	Ounces	=	373.2420	Grams

D. Kitchen Weights and Measures

1 U.S. Pint	= 16	Fluid Ounces
1 Standard Cup	= 8	Fluid Ounces
1 Tablespoon	= 0.5	Fluid Ounces (15 Cubic Centimeters)
1 Teaspoon	= 0.16	Fluid Ounces (5 Cubic Centimeters)

 **METRIC SYSTEM**

Prefixes

A. Mega	= 1000000	E. Deci	= 0.1	
B. Kilo	= 1000	F. Centi	= 0.01	
C. Hecto	= 100	G. Milli	= 0.001	
D. Deka	= 10	H. Micro	= 0.000001	

Linear Measure

(The Unit is the Meter = 39.37 Inches)

1 Centimeter	=	10	Millimeters	= 0.3937011	Inch
1 Decimeter	=	10	Centimeters	= 3.9370113	Inches
1 Meter	=	10	Decimeters	= 1.0936143	Yards
				= 3.2808429	Feet
1 Dekameter	=	10	Meters	= 10.936143	Yards
1 Hectometer	=	10	Dekameters	= 109.36143	Yards
1 Kilometer	=	10	Hectometers	= 0.62137	Mile
1 Myriameter	=	10000	Meters		

Square Measure

(The Unit is the Square Meter = 1549.9969 SQ. Inches)

1 SQ. Centimeter	=	100 SQ. Millimeters	=	0.1550	Square Inch
1 SQ. Decimeter	=	100 SQ. Centimeters	=	15.550	Square Inches
1 SQ. Meter	=	100 SQ. Decimeters	=	10.7639	Square Feet
1 SQ. Dekameter	=	100 SQ. Meters	=	119.60	Square Yards
1 SQ. Hectometer	=	100 SQ. Dekameters			
1 SQ. Kilometer	=	100 SQ. Hectometers			

(The Unit is the "Are" = 100 SQ. Meters)

1 Centiare	=	10	Milliares	= 10.7643	Square Feet
1 Deciare	=	10	Centiares	= 11.96033	Square Yards
1 Are	=	10	Deciares	= 119.6033	Square Yards
1 Decare	=	10	Ares	= 0.247110	Acres
1 Hectare	=	10	Decares	= 2.471098	Acres
1 SQ. Kilometer	=	100	Hectares	= 0.38611	Square Mile

Cubic Measure

(The Unit is the "Stere" = 61025.38659 CU. INs.)

1 Decistere	= 10 Centisteres	= 3.531562	Cubic Foot
1 Stere	= 10 Decisteres	= 1.307986	Cubic Yards
1 Dekastere	= 10 Steres	= 13.07986	Cubic Yards

 **METRIC SYSTEM**

Cubic Measure *(continued)*

(The Unit is the Meter = 39.37 Inches)

1 CU. Centimeter	= 1000 CU. Millimeters	=	0.06102 Cubic Inches
1 CU. Decimeter	= 1000 CU. Centimeters	=	61.02374 Cubic Inches
1 CU. Meter	= 1000 CU. Decimeters	=	35.31467 Cubic Feet
	= 1 Stere	=	1.30795 Cubic Yards
1 CU. Centimeter (Water)		=	1 Gram
1000 CU. Centimeters (Water) = 1 Liter		=	1 Kilogram
1 CU. Meter (1000 Liters)		=	1 Metric Ton

Measures of Weight

(The Unit is the Gram = 0.035274 Ounces)

1 Milligram	=		=	0.015432	Grains
1 Centigram	=	10 Milligrams	=	0.15432	Grains
1 Decigram	=	10 Centigrams	=	1.5432	Grains
1 Gram	=	10 Decigrams	=	15.4323	Grains
1 Dekagram	=	10 Grams	=	5.6438	Drams
1 Hectogram	=	10 Dekagrams	=	3.5274	Ounces
1 Kilogram	=	10 Hectograms	=	2.2046223	Pounds
1 Myriagram	=	10 Kilograms	=	22.046223	Pounds
1 Quintal	=	10 Myriagrams	=	1.986412	Hundredweight
1 Metric Ton	=	10 Quintal	=	22045.622	Pounds
1 Gram	=	0.56438 Drams			
1 Dram	=	1.77186 Grams			
	=	27.3438 Grains			
1 Metric Ton	=	2204.6223 Pounds			

Measures of Capacity

(The Unit is the Liter = 1.0567 Liquid Quarts)

1 Centiliter	=	10 Milliliters	=	0.338	Fluid Ounces
1 Deciliter	=	10 Centiliters	=	3.38	Fluid Ounces
1 Liter	=	10 Deciliters	=	33.8	Fluid Ounces
1 Dekaliter	=	10 Liters	=	0.284	Bushel
1 Hectoliter	=	10 Dekaliters	=	2.84	Bushels
1 Kiloliter	=	10 Hectoliters	=	264.2	Gallons

Note: $\dfrac{\text{Kilometers}}{8} \times 5 = \text{Miles}$ or $\dfrac{\text{Miles}}{5} \times 8 = \text{Kilometers}$

 **METRIC SYSTEM**

Metric Designator and Trade Sizes

Metric Designator

12	16	21	27	35	41	53	63	78	91	103	129	155
⅜	½	¾	1	1¼	1½	2	2½	3	3½	4	5	6

Trade Size

Source: NFPA 70, *National Electrical Code®*, NFPA, Quincy, MA, 2023, Table 300.1(C), as modified.

U.S. Weights and Measures/Metric Equivalent Chart

	In.	Ft	Yd.	Mile	mm	cm	m	km
1 Inch =	1	.0833	.0278	1.578×10^{-5}	25.4	2.54	.0254	2.54×10^{-5}
1 Foot =	12	1	.333	1.894×10^{-4}	304.8	30.48	.3048	3.048×10^{-4}
1 Yard =	36	3	1	5.6818×10^{-4}	914.4	91.44	.9144	9.144×10^{-4}
1 Mile =	63360	5280	1760	1	1609344	160934.4	1609.344	1609344
1 mm =	.03937	.00032808	1.0936×10^{-3}	6.2137×10^{-7}	1	0.1	0.001	0.000001
1 cm =	.3937	.0328084	.0109361	6.2137×10^{-6}	10	1	0.01	0.00001
1 m =	39.37	3.280.84	1.093.61	6.2137×10^{-4}	1000	100	1	0.001
1 km =	39370	3280.84	1093.61	0.62137	1000000	100000	1000	1

in. = inches Ft = foot Yd. = yard mm = millimeter cm = centimeter m = meter km = kilometer

Explanation of Scientific Notation

Scientific notation (powers of 10) is simply a way of expressing very large or very small numbers in a more compact format. Any number can be expressed as a number between 1 and 10, multiplied by a power of 10 (which indicates the correct position of the decimal point in the original number). Numbers greater than 10 have positive powers of 10, and numbers less than 1 have negative powers of 10.

Example: $186000 = 1.86 \times 10^5$ $0.000524 = 5.24 \times 10^{-4}$

Useful Conversions/Equivalents

1 BTU	Raises 1 lb of water 1°F
1 Gram Calorie	Raises 1 gram of water 1°C
1 Circular Mil	Equals 0.7854 sq. mil
1 SQ. Mil	Equals 1.2732 cir. mils
1 Mil	Equals 0.001 in.

To determine circular mil (cmil) of a conductor:

Round Conductor . . . .   cmil = (Diameter in mils)^2

Rectangle Bus Bar . . . . $\text{cmil} = \dfrac{\text{Width} \times \text{Thickness} \times 1{,}000{,}000}{0.7854}$

Notes: 1 millimeter = 39.37 mils   1 cir. millimeter = 1550 cir. mils
1 sq. millimeter = 1974 cir. mils

DECIMAL EQUIVALENTS

Fraction					Decimal	Fraction					Decimal
1/64					.0156	33/64					.5156
2/64	1/32				.0313	34/64	17/32				.5313
3/64					.0469	35/64					.5469
4/64	2/32	1/16			.0625	36/64	18/32	9/16			.5625
5/64					.0781	37/64					.5781
6/64	3/32				.0938	38/64	19/32				.5938
7/64					.1094	39/64					.6094
8/64	4/32	2/16	1/8		.125	40/64	20/32	10/16	5/8		.625
9/64					.1406	41/64					.6406
10/64	5/32				.1563	42/64	21/32				.6563
11/64					.1719	43/64					.6719
12/64	6/32	3/16			.1875	44/64	22/32	11/16			6875
13/64					.2031	45/64					.7031
14/64	7/32				.2188	46/64	23/32				.7188
15/64					.2344	47/64					.7344
16/64	8/32	4/16	2/8	1/4	.25	48/64	24/32	12/16	6/8	3/4	.75
17/64					.2656	49/64					.7656
18/64	9/32				.2813	50/64	25/32				.7813
19/64					.2969	51/64					.7969
20/64	10/32	5/16			.3125	52/64	26/32	13/16			.8125
21/64					.3281	53/64					.8281
22/64	11/32				.3438	54/64	27/32				.8438
23/64					.3594	55/64					.8594
24/64	12/32	6/16	3/8		.375	56/64	28/32	14/16	7/8		.875
25/64					.3906	57/64					.8906
26/64	13/32				.4063	58/64	29/32				.9063
27/64					.4219	59/64					.9219
28/64	14/32	7/16			.4375	60/64	30/32	15/16			.9375
29/64					.4531	61/64					.9531
30/64	15/32				.4688	62/64	31/32				.9688
31/64					.4844	63/64					9844
32/64	1632	8/16	48	2/4	.5	64/64	32/32	16/16	8/8	4/4	1.000

Decimals are rounded to the nearest 10000th.

TWO-WAY CONVERSION TABLE

To convert from the unit of measure in Column B to the unit of measure in Column C, multiply the number of units in Column B by the multiplier in Column A. To convert from Column C to B, use the multiplier in Column D.

Example: To convert 1000 BTUs to Calories, find the "BTU - Calorie" combination in Columns B and C. "BTU" is in Column B and "Calorie" is in Column C; so we are converting from B to C. Therefore, we use Column A multiplier. 1000 BTUs x 251.996 = 251996 Calories.

To convert 251996 Calories to BTUs, use the same "BTU. Calorie" combination. But this time you are converting from C to B. Therefore, use Column D multiplier. 251996 Calories x .0039683 = 1000 BTUs.

$$A \times B = C \qquad\qquad \& \qquad\qquad D \times C = B$$

To Convert from B to C, Multiply B x A:			To Convert from C to B, Multiply C x D:
A	**B**	**C**	**D**
43560	Acre	Sq. foot	2.2956×10^{-5}
1.5625×10^{-3}	Acre	Sq. mile	640
6,4516	Ampere per sq. cm.	Ampere per sq. in.	0.155003
1.256637	Ampere (turn)	Gilberts	0.79578
33.89854	Atmosphere	Foot of H_2O	0.029499
29.92125	Atmosphere	Inch of Hg	0.033421
14.69595	Atmosphere	Pound force/sq. In.	0.06804
251.996	BTU	Calorie	3.96832×10^{-3}
778.169	BTU	Foot-pound force	1.28507×10^{-3}
3.93015×10^{-4}	BTU	Horsepower·hour	2544.43
1055.056	BTU	Joule	9.47817×10^{-4}
2.9307×10^{-4}	BTU	Kilowatt·hour	3412.14
3.93015×10^{-4}	BTU/hour	Horsepower	2544.43
$2,93071 \times 10^{-4}$	BTU/hour	Kilowatt	3412.1412
0.293071	BTU/hour	Wan	3.41214
4.19993	BTU/minute	Calorie/second	0.23809
0.0235809	BTU/minute	Horsepower	42.4072
17.5843	BTU/minute	Walt	0.0568

(continued on next page)

TWO-WAY CONVERSION TABLE

A	B	C	D
4.1868	Calorie	Joule	0.238846
0.0328084	Centimeter	Foot	30.48
0.3937	Centimeter	Inch	2.54
0.00001	Centimeter	Kilometer	100000
0.01	Centimeter	Meter	100
6.2137×10^{-6}	Centimeter	Mile	160934.4
10	Centimeter	Millimeter	0.1
0.010936	Centimeter	Yard	91.44
7.85398×10^{-7}	Circular mil	Sq. inch	1.273239×10^{-6}
0.000507	Circular mil	Sq. millimeter	1973.525
0.06102374	Cubic centimeter	Cubic inch	16.387065
0.028317	Cubic foot	Cubic meter	35.31467
1.0197×10^{-3}	Dyne	Gram force	980.665
1×10^{-5}	Dyne	Newton	100000
1	Dyne centimeter	Erg	1
7.376×10^{-8}	Erg	Foot pound force	1.355818×10^{7}
2.777×10^{-14}	Erg	Kilowatthour	3.6×10^{13}
1.0×10^{-7}	Erg/second	Watt	1.0×10^{7}
12	Foot	Inch	0.0833
3.048×10^{-4}	Foot	Kilometer	3280.84
0.3048	Foot	Meter	3.28084
1.894×10^{-4}	Foot	Mile	5280
304.8	Foot	Millimeter	0.00328
0.333	Foot	Yard	3
10.7639	Foot candle	Lux	0.0929
0.882671	Foot of H_2O	Inch of Hg	1.13292
5.0505×10^{-7}	Foot pound force	Horsepower-hour	1.98×10^{6}
1.35582	Foot pound force	Joule	0.737562
3.76616×10^{-7}	Foot pound force	Kilowatthour	2.655223×10^{6}
3.76616×10^{-4}	Foot pound force	Watt-hour	2655.22
3.76616×10^{-7}	Foot pound force/hour	Kilowatt	2.6552×10^{6}
3.0303×10^{-5}	Foot pound force/minute	Horsepower	33000

TWO-WAY CONVERSION TABLE

	To Convert from B to C, Multiply B x A:		To Convert from C to B, Multiply C x D:
A	**B**	**C**	**D**
2.2597×10^{-5}	Foot pnd. force/minute	Kilowatt	44253.7
0.022597	Foot pnd. force/minute	Watt	44.2537
1.81818×10^{-3}	Foot pnd. force/second	Horsepower	550
1.355818×10^{-3}	Foot pnd. force/second	Kilowatt	737.562
0.7457	Horsepower	Kilowatt	1.34102
745.7	Horsepower	Watt	0.00134
0.0022046	Gram	Pound mass	453.592
2.54×10^{-5}	Inch	Kilometer	39370
0.0254	Inc It	Meter	39.37
1.578×10^{-5}	Inch	Mile	63360
25.4	Inch	Millimeter	0.03937
0.0278	Inch	Yard	36
0.07355	Inch of H.0	Inch ol Hg	13.5951
2.7777×10^{-7}	Joule	Kilowatthour	3.6×10^{6}
2.7777×10^{-4}	Joule	Watt hour	3600
1	Joule	Watt second	1
1000	Kilometer	Meter	0.001
0.62137	Kilometer	Mile	1.609344
1000000	Kilometer	Millimeter	0.000001
1093.61	Kilometer	Yard	9.144×10^{-4}
0.000621	Meter	Mile	1609.344
1000	Meter	Millimeter	0.001
1.0936	Meter	Yard	0.9144
1609344	Mile	Millimeter	6.2137×10^{-7}
1760	Mile	Yard	5.681×10^{-4}
1.0936×10^{-3}	Millimeter	Yard	914.4
0.224809	Newton	Pound force	4.44822
0.03108	Pound	Slug	32.174
0.0005	Pound	Ton (short)	2000
0.155	Sq. centimeter	Sq. inch	6.4516
0.092903	Sq. foot	Sq. meter	10.76391
0.386102	Sq. kilometer	Sq. mile	2.589988

METALS

Metal	SYMB	Spec. Grav.	Melt Point		Elec.Cond. %Copper	Pounds Cu Feet
			c°	F°		
Aluminum	Al	2.71	660	1220	64.9	0978
Antimony	Sb	6.62	630	1167	4.42	.2390
Arsenic	As	5.73	–	–	4,9	.2070
Beryllium	Be	1.83	1280	2336	9.32	.0660
Bismuth	BI	9.80	271	520	1.50	.3540
Brass (70-30)		8.51	900	1652	28.0	.3070
Bronze (5% SN)		8.87	1000	1832	18,0	.3200
Cadmium	Cd	8.65	321	610	22.7	.3120
Calcium	Ca	1.55	850	1562	50.1	.0560
Cobalt	Co	8.90	1495	2723	17,8	.3210
Copoer	Cu					
Rolled		8.89	1083	1981	100.0	.3210
Tubing		8.95	–	–	100.0	.3230
Gold	Au	19.30	1063	1945	71.2	.6970
Graphite		2.25	3500	6332	10^{-3}	.0812
Indium	In	7.30	156	311	20.6	.2640
Iridium	Ir	22.40	2450	4442	32.5	.8090
Iron	Fe	7.20	1200–1400	2192–2552	17,6	.2600
Malleable		7.20	1500–1600	2732–2912	10	.2600
Wrought		7.70	1500–1600	2732–2912	10	.2780
Lead	Pb	11.40	327	621	8.35	.4120
Magnesium	Mg	1.74	651	1204	38.7	.0628

146

⌂ METALS

Metal	SYMB	Spec. Grav.	Melt Point C°	Melt Point F°	Elec.Cond. %Copper	LBS. CU."
Manganese	Mn	7.20	1245	2273	0.9	.2600
Mercury	Hg	13.65	−38.9	-37.7	1,80	.4930
Molybdenum	Mo	10.20	2620	4748	36.1	.3680
Monel (63 - 37)		8.87	1300	2372	3,0	.3200
Nickel	Ni	8.90	1452	2646	25.0	.3210
Phosphorous	P	1.82	44.1	111.4	10^{-17}	.0657
Platinum	Pt	21346	1773	3221	17.5	.7750
Potassium	K	0.860	62.3	144.1	28	.0310
Selenium	Se	4.81	220	428	14.4	.1740
Silicon	Si	2,40	1420	2588	10^{-5}	.0866
Silver	Ag	10.50	960	1760	106	.3790
Steel (Carbon)		7.84	1330–1380	2436–2516	10	.2830
Stainless						
(18-8)		7.92	1500	2732	2.5	.2860
(13-CR)		7.78	1520	2768	3.5	.2810
Tantalum	Ta	16.60	2900	5414	13.9	.599
Tellurium	Te	6.20	450	846	10^{-5}	.224
Thorium	Th	11.70	1845	3353	9.10	.422
Tin	Sn	7.30	232	449	15.00	.264
Titanium	Tl	4,50	1800	3272	2.10	.162
Tungsten	W	19.30	3410	-	31.50	.697
Uranium	U	18.70	1130	2066	2.80	.675
Vanadium	V	5.96	1710	3110	6.63	.215
Zinc	Zn	7.14	419	786	29.10	.258
Zirconium	Zr	6.40	1700	3092	4.20	.231

METALS

Specific Resistance (k)

The specific resistance (K) of a material is the resistance offered by a wire of this material that is 1 foot long with a diameter of 1 mil.

Material	"K"	Material	"K"
Brass	43.0	Aluminum	17.0
Constantan	295	Monel	253
Copper	10.8	Nichrome	600
German Silver 18%	200	Nickel	947
Gold	14.7	Tantalum	93.3
Iron (Pure)	60.0	Tin	69.0
Magnesium	276	Tungsten	34.0
Manganin	265	Silver	9.7

Note: 1. The resistance of a wire is directly proportional to the specific resistance of the material

2. "K" = Specific Resistance

3. Resistance varies with temperature. See *NEC* Chapter 9, Table 8, Note 1.

CENTIGRADE AND FAHRENHEIT THERMOMETER SCALES

°C	°F	°C	°F	°C	°F	°C	°F
0	32						
1	33.8	26	78.8	51	123.8	76	168,8
2	35.6	27	80.6	52	125.6	77	170.6
3	37.4	28	82.4	53	127.4	78	172.4
4	39.2	29	84.2	54	129.2	79	174.2
5	41	30	86	55	131	80	176
6	42.8	31	87.8	56	132.8	81	177.8
7	44.6	32	89.6	57	134.6	82	179.6
8	46.4	33	91.4	58	136.4	83	181.4
9	48.2	34	93.2	59	138.2	84	183.2
10	50	35	95	60	140	85	185
11	51.8	36	96.8	61	141.8	86	186.8
12	53.6	37	98.6	62	143.6	87	188.6
13	55.4	38	100.4	63	145.4	88	190.4
14	57.2	39	102.2	64	147.2	89	192.2
15	59	40	104	65	149	90	194
16	60.8	41	105.8	66	150.8	91	195.8
17	62.6	42	107.6	67	152.6	92	197.6
18	64.4	43	109.4	68	154.4	93	199.4
19	66.2	44	111.2	69	156.2	94	201.2
20	68	45	113	70	158	95	203
21	69.8	46	114.8	71	159.8	96	204.8
22	71.6	47	116.6	72	161.6	97	206.6
23	73.4	48	118.4	73	163.4	98	208.4
24	75.2	49	120.2	74	165.2	99	210.2
25	77	50	122	75	167	100	212

1. Temp. °C $= \frac{5}{9}$ x (Temp. °F − 32)

2. Temp. °F $= (\frac{9}{5}$ x Temp. °C) + 32

3. Ambient temperature is the temperature of the surrounding cooling medium.

4. Rated temperature rise is the permissible rise in temperature above ambient when operating under load.

USEFUL MATH FORMULAS

Right triangle

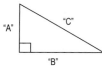

$A = \sqrt{C^2 - B^2}$

$B = \sqrt{C^2 - A^2}$

$C = \sqrt{A^2 - B^2}$

Area $= 0.5 \times (A \times B)$

Oblique triangle

Solve as two right triangles

Sphere

Surface Area $= D^2 \times 3.1416$
Volume $= D^3 \times 0.5236$

Cylindrical

Volume $=$
Area of end
x height

Cone

Volume $=$
Area of end x height $\div 3$

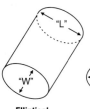

Elliptical

Solve the same
as cylindrical

Ellipse

Area $= a \times b \times 3.1416$
(assuming C is center)

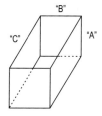

Rectangular prism

Volume $= A \times B \times C$
Area $= 2(AB + AC + BC)$

150

 **THE CIRCLE**

Definition: A closed curve in a plane having every point an equal distance from a fixed point called the center.

Circumference: The distance around a circle
Diameter: The distance across a circle through the center
Radius: The distance from the center to the edge of a circle
ARC: A part of the circumference
Chord: A straight line connecting the ends of an arc
Segment: An area bounded by an arc and a chord
Sector: A part of a circle enclosed by two radii and the arc that they cutoff (like a slice of pie)

Circumference of a Circle = 3.1416 x 2 x radius
Area of a Circle = 3.1416 x radius^2
ARC Length = Degrees In arc x radius x 0.01745
Radius Length = One-half length of diameter
Sector Area = One-half length of arc x radius
Chord Length = $2 \sqrt{A \times B}$
Segment Area = Sector area minus triangle area

Note:
3.1416 x 2 x R = 360°,
or 0.0087266 x 2 x R = 1°, or
0.01745 x R = 1°
This gives us the arc formula.
Degrees x Radius x 0.01745 =
Developed Length

Example:
For a 90° conduit bend, having
a radius of 17.25":
90 x 17.25" x 0.01745 =
Developed Length
27,09" = Developed Length

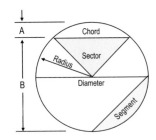

FRACTIONS

Definitions

A. A <u>fraction</u> is a quantity less than a unit.

B. A <u>numerator</u> is the term of a fraction indicating how many of the parts of a unit are to be taken. In a common fraction, it appears above or to the left of the line.

C. A <u>denominator</u> is the term of a fraction indicating the number of equal parts into which the unit is divided. In a common fraction, it appears below or to the right of the line.

D. **_Examples:_**

$$(1.) \quad \frac{1}{2} = \frac{\text{Numerator}}{\text{Denominator}} = \text{Fraction}$$

$$(2.) \quad \text{Numerator} \longrightarrow \frac{1}{2} \longleftarrow \text{Denominator}$$

To Add or Subtract

To solve: $\frac{1}{2} - \frac{2}{3} + \frac{3}{4} - \frac{5}{6} + \frac{7}{12} = ?$

A. Determine the lowest common denominator that each of the denominators 2, 3, 4, 6, and 12 will divide into an even number of times.

The lowest common denominator is 12.

B. Work one fraction at a time using the formula:

Common Denominator		
Denominator of Fraction	x	**Numerator of Fraction**

(1.) $\frac{12}{2} \times 1 = 6 \times 1 = 6$ $\frac{1}{2}$ becomes $\frac{6}{12}$

(2.) $\frac{12}{3} \times 2 = 4 \times 2 = 8$ $\frac{2}{3}$ becomes $\frac{8}{12}$

(3.) $\frac{12}{4} \times 3 = 3 \times 3 = 9$ $\frac{3}{4}$ becomes $\frac{9}{12}$

(4.) $\frac{12}{6} \times 5 = 2 \times 5 = 10$ $\frac{5}{6}$ becomes $\frac{10}{12}$

(5.) $\frac{7}{12}$ remains $\frac{7}{12}$

⬚ FRACTIONS

To Add or Subtract *(continued)*

C. We can now convert the problem from Its original form to its new form using 12 as the common denominator.

$$1/2 - 2/3 + 3/4 - 5/6 + 7/12 \qquad = \text{Original form}$$

$$\frac{6 - 8 + 9 - 10 + 7}{12} \qquad = \text{Present form}$$

$$\frac{4}{12} \; = \; \frac{1}{3} \qquad \text{Reduced to lowest form}$$

D. To convert fractions to decimal form, simply divide the numerator of the fraction by the denominator of the fraction.

   Example: $\frac{1}{3}$ $= 1$ Divided by $3 \; = 0.333$

To Multiply

A. The numerator of fraction # 1 times the numerator of fraction #2 is equal to the numerator of the product.

B. The denominator of fraction #1 times the denominator of fraction #2 is equal to the denominator of the product.

C. ***Example:***

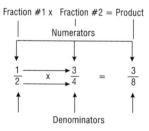

Note: To change ⅜ to decimal form, divide 3 by 8 $= 0.375$

To Divide

A. The numerator of fraction #1 times the denominator of fraction #2 is equal to the numerator of the quotient.

B. The denominator of fraction #1 times the numerator of fraction #2 is equal to the denominator of the quotient.

C. **Example:** $\dfrac{1}{2} \div \dfrac{3}{4}$

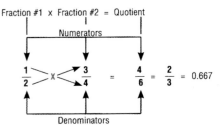

D. An alternate method for dividing by a fraction is to multiply by the reciprocal of the divisor (the second fraction in a division problem).

E. **Example:** $\dfrac{1}{2} \div \dfrac{3}{4}$

The reciprocal of $\dfrac{3}{4}$ is $\dfrac{4}{3}$

so, $\dfrac{1}{2} \div \dfrac{3}{4} = \dfrac{1}{2} \times \dfrac{4}{3} = \dfrac{4}{6} = \dfrac{2}{3} = 0.667$

⚓ EQUATIONS

The word "Equation" means equal or the same as.

Example: $\qquad 2 \times 10 = 4 \times 5$

$\qquad\qquad\qquad 20 = 20$

Rules

A. **The same number may be added to both sides of an equation without changing its values.**

Example: $\qquad (2 \times 10) + 3 = (4 \times 5) + 3$

$\qquad\qquad\qquad 23 = 23$

B. **The same number may be subtracted from both sides of an equation without changing its values.**

Example: $\qquad (2 \times 10) - 3 = (4 \times 5) - 3$

$\qquad\qquad\qquad 17 = 17$

C. **Both sides of an equation may be divided by the same number without changing its values.**

Example: $\qquad \dfrac{2 \times 10}{20} = \dfrac{4 \times 5}{20}$

$\qquad\qquad\qquad 1 = 1$

D. **Both sides of an equation may be multiplied by the same number without changing its values.**

Example: $\qquad 3 \times (2 \times 10) = 3 \times (4 \times 5)$

$\qquad\qquad\qquad 60 = 60$

E. **Transposition:**

The process of moving a quantity from one side of an equation to the other side of an equation by changing its sign of operation.

1. **A term may be transposed if its sign is changed from plus (+) to minus (−), or from minus (−) to plus (+).**

Example: $\quad X + 5 = 25$

$\qquad\qquad X + 5 - 5 = 25 - 5$

$\qquad\qquad X = 20$

⚡ EQUATIONS

E. **Transposition** *(continued)*:

2. **A multiplier may be removed from one side of an equation by making it a divisor on the other side; or a divisor may be removed from one side of an equation by making it a multiplier on the other side.**

Example: Multiplier from one side of equation (4) becomes divisor on other side.

$$4X = 40 \text{ becomes } X = \frac{40}{4} = 10$$

Example: Divisor from one side of equation becomes multiplier on other side.

$$\frac{X}{4} = 10 \text{ becomes } X = 10 \times 4$$

Signs

A. **Addition** of numbers with *different* signs:

1. **Rule: Use the sign of the larger and subtract.**

Example:

$$\begin{array}{r} +3 \\ +\ -2 \\ \hline +1 \end{array} \qquad \begin{array}{r} -2 \\ +\ +3 \\ \hline +1 \end{array}$$

B. **Addition** of numbers with the *same* signs:

2. **Rule: Use the common sign and add.**

Example:

$$\begin{array}{r} +3 \\ +\ +2 \\ \hline +5 \end{array} \qquad \begin{array}{r} -3 \\ +\ -2 \\ \hline -5 \end{array}$$

C. **Subtraction** of numbers with *different* signs:

3. **Rule: Change the sign of the subtrahend (the second number in a subtraction problem) and proceed as in addition.**

Example:

$$\begin{array}{r} +3 \\ -\ -2 \\ \hline \end{array} = \begin{array}{r} +3 \\ +\ +2 \\ \hline +5 \end{array} \qquad \begin{array}{r} -2 \\ -\ +3 \\ \hline \end{array} = \begin{array}{r} -2 \\ +\ -3 \\ \hline -5 \end{array}$$

EQUATIONS

Signs *(continued)*

D. **Subtraction** of numbers with the *same* signs:

    4. **Rule:** Change the sign of the subtrahend (the second number in a subtraction problem) and proceed as in addition.

Example:

$$\begin{array}{c} +3 \\ -\ +2 \end{array} = \begin{array}{c} +3 \\ +\ -2 \\ \hline +1 \end{array} \qquad \begin{array}{c} -3 \\ -\ -2 \end{array} = \begin{array}{c} -3 \\ +\ +2 \\ \hline -1 \end{array}$$

E. **Multiplication:**

    5. **Rule:** The product of any two numbers having <u>like</u> signs is <u>positive</u>. The product of any two numbers having <u>unlike</u> signs is <u>negative</u>.

Example:
$(+3) \times (-2) = -6$
$(-3) \times (+2) = -6$
$(+3) \times (+2) = +6$
$(-3) \times (-2) = +6$

F. **Division:**

    6. **Rule:** If the divisor and the dividend have <u>like</u> signs, the sign of the quotient is <u>positive</u>. If the divisor and dividend have <u>unlike</u> signs, the sign of the quotient is <u>negative</u>.

Example:

$$\frac{+6}{-2} = -3 \qquad \frac{+6}{+2} = +3$$

$$\frac{-6}{+2} = -3 \qquad \frac{-6}{-2} = +3$$

NATURAL TRIGONOMETRIC FUNCTIONS

Angle	Sine	Cosine	Tangent	Cotangent	Secant	Cosecant	
0	.0000	1.0000	.0000		1.0000		90
1	.0175	.9998	.0175	57.2900	1.0002	57.2987	89
2	.0349	.9994	.0349	28.6363	1.0006	28.6537	88
3	.0523	.9986	.0524	19.0811	1.0014	19.1073	87
4	.0698	.9976	.0699	14.3007	1.0024	14.3356	86
5	.0872	.9962	.0875	11.4301	1.0038	11.4737	85
6	.1045	.9945	.1051	9.5144	1.0055	9.5668	84
7	.1219	.9925	.1228	8.1443	1.0075	8.2055	83
8	.1392	.9903	.1405	7.1154	1 0098	7.1853	82
9	.1564	.9877	.1584	6.3138	1.0125	6.3925	81
10	.1736	.9848	.1763	5.6713	1.0154	5.7588	80
11	.1908	.9816	.1944	5.1446	1.0187	5.2408	79
12	.2079	.9781	.2126	4.7046	1.0223	4 8097	78
13	.2250	.9744	.2309	4.3315	1.0263	4.4454	77
14	.2419	.9703	.2493	4.0108	1.0306	4.1336	76
15	2588	.9659	.2679	3.7321	1.0353	3 8637	75
16	.2756	.9613	.2867	3.4874	1.0403	3.6280	74
17	.2924	.9563	.3057	3.2709	1.0457	3.4203	73
18	.3090	.9511	.3249	3.0777	1.0515	3.2361	72
19	.3256	.9455	.3443	2.9042	1.0576	3.0716	71
20	.3420	.9397	.3640	2.7475	1.0642	2.9238	70
21	.3584	.9336	.3839	26051	1.0711	2.7904	69
22	.3746	.9272	.4040	2.4751	1.0785	2.6695	68
23	.3907	.9205	.4245	2.3559	1.0864	2.5593	67
24	.4067	.9135	.4452	2.2460	1.0946	2.4586	66
25	.4226	.9063	.4663	2.1445	1.1034	2.3662	65
26	.4384	.8988	.4877	2.0503	1.1126	2.2812	64
27	.4540	.8910	.5095	1.9626	1.1223	2.2027	63
	Cosine	Sim	Cotangt.	Tangent	Cosecant	Secant	Angle

(continued on next page)

NATURAL TRIGONOMETRIC FUNCTIONS

Angle	Sine	Cosine	Tangent	Cotangent	Secant	Cosecant	
28	.4695	.8829	.5317	1.8807	1.1326	2.1301	62
29	.4848	.8746	.5543	1.8040	1.1434	2.0627	61
30	.5000	.8660	.5774	1.7321	1.1547	2.0000	60
31	.5150	.8572	.6009	1 6643	1.1666	1.9416	59
32	.5299	.8480	.6249	1.6003	1.1792	1.8871	58
33	.5446	.8387	.6494	1.5399	1.1924	1.8361	57
34	.5592	.8290	.6745	1.4826	1.2062	1.7883	56
35	.5736	.8192	7002	1.4281	1.2208	1.7434	55
36	.5878	.8090	.7265	1.3764	1.2361	1.7013	54
37	.6018	.7986	.7536	1.3270	1.2521	1.6616	53
38	.6157	.7880	.7813	1.2799	1.2690	1.6243	52
39	.6293	.7771	.8098	1.2349	1.2868	1.5890	51
40	.6428	.7660	.8391	1.1918	1.3054	1.5557	50
41	.6561	.7547	.8693	1.1504	1.3250	1.5243	49
42	.6691	.7431	.9004	1.1106	1.3456	1.4945	48
43	.6820	.7314	.9325	1.0724	1.3873	1.4663	47
44	.6947	.7193	9657	1.0355	1.3902	1.4396	48
45	,7071	.7071	1.0000	1.0000	1.4142	1.4142	45
	Coalne	Sine	Cotingt.	Tengint	Cosecant	Secant	Angle

Note: For angles (0-45, use top row and left column.

For angles 45-90, use bottom row and right column.

⌐ TRIGONOMETRY

Trigonometry is the mathematics dealing with the relations of sides and angler oftnangles.

A triangle is a figure enclosed by three straight sides. The sum of the three angles is 180°. All triangles have six parts: three angles and three sides opposite the angles.

Right Triangles are triangles that have one angle of 90° and two angles of less than 90°.

To help you remember the six trigonometric functions, memorize:

"Oh Hell Another Hour of Andy"

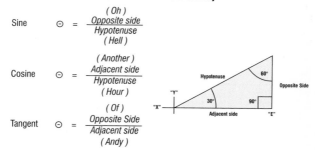

		(Oh)
Sine	⊖ =	$\dfrac{\text{Opposite side}}{\text{Hypotenuse}}$
		(Hell)

		(Another)
Cosine	⊖ =	$\dfrac{\text{Adjacent side}}{\text{Hypotenuse}}$
		(Hour)

		(Of)
Tangent	⊖ =	$\dfrac{\text{Opposite Side}}{\text{Adjacent side}}$
		(Andy)

Now, use backward: **"Andy of Hour Another Hell Oh"**

		(Andy)	
Cotangent	θ =	$\dfrac{\text{Adjacent side}}{\text{Opposite side}}$	Always place the angle to be solved at the vertex (where "X" and "Y" cross).
		(Of)	
		(Hour)	
Secant	θ	$\dfrac{\text{Hypotenuse}}{\text{Adjacent side}}$	
		(Another)	
		(Hell)	
Cosecant	θ	$\dfrac{\text{Hypotenuse}}{\text{Opposite side}}$	*Note:*
		(Oh)	θ = Theta = Any Angle

⑤ BENDING OFFSETS WITH TRIGONOMETRY

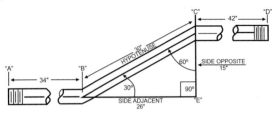

The Cosecant of the Angle Times the Offset Desired Is Equal to the Distance Between the Centers of the Bends.

Example

To make a fifteen inch (15") offset, using thirty (30) degree bends:

1. Use Trig. Table (pages 158–159) to find the Cosecant of a thirty (30) degree angle. We find it to be two (2).
2. Multiply two (2) times the offset desired, which is fifteen (15) inches to determine the distance between bend "B" and bend "C." The answer is thirty (30) inches.

To mark the conduit for bending:

1. Measure from end of Conduit "A" thirty-four (34) inches to center of first bend "B," and mark.
2. Measure from mark "B" thirty (30) inches to center of second bend "C" and mark.
3. Measure from mark "C" forty-two (42) inches to "D," and mark. Cut, ream, and thread conduit before bending.

Rolling Offsets

To determine how much offset is needed to make a rolling offset:

1. Measure vertical required. Use work table (any square will do) and measure from corner this amount and mark.
2. Measure horizontal required. Measure 90° from the vertical line measurement (starting in same corner) and mark.
3. The diagonal distance between these marks will be the amount of offset required.

Note: Shrink is hypotenuse minus the side adjacent.

161

 **ONE SHOT BENDS**

Shrink Constant for Angles Less Than 60° = Angle/120

Example: The shrink constant for 45° is 3/8"

$$45/120 = 3/8"$$

Shrink Constant for 60° to 90° Angles = Angle/100

Example: The shrink constant for 45° is 3/8"

$$45/100 = 3/8"$$

Multiplier = (60/Angle) + (Angle/200) - 0.15

Example: The Multiplier for 50° is 1.3.

$$(60/50) + (50/200) - 0.15 = 1.3$$

The calculation for this multiplier is an error of less than half a percent.

Bend Length = (Angle x D)/60

Example: If putting a 40° bend in 3/4" conduit, the bend length is 4".

$$(40 \times 6")/60 = 4"$$

"D" is the deduct for whatever size conduit is being run. This formula works for any angle between 0 and 90°.

Note: With these formulas, the entire run in pieces (straight and curved) can be seen, including exactly where each piece starts and where it ends. This allows the bender direction (hook facing east or west) to be chosen at each point in the run, and bend marks can be laid out accordingly

⚡ CHICAGO-TYPE BENDERS: 90° BENDING

"A" to "C" = Stub-Up
"C" to "D" = Tail
"C" = Back of Stub-Up
"C" = Bottom of Conduit

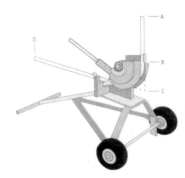

Note:
There are many variations
of this type bender, but
most manufacturers
offer two sizes.
The *small* size shoe takes
1/2", 3/4", and 1" conduit.
The *large* size shoe takes
1¼" and 1½" conduit.

To determine the "take-up" and "shrink" of each
size conduit for a particular bender to make 90°
bends:
1. Use a straight piece of scrap conduit.
2. Measure exact length of scrap conduit, "A" to "D".

"D"   Oringinal Measurement   "A"

3. Place conduit in bender. Mark at edge of shoe, "B".
4. Level conduit. Bend ninety, and count number of pumps. Be sure to
   keep notes on each size conduit used.
5. After bending ninety:
   A. Distance between "B" and "C" is the Take-Up.
   B. Original measurement of the scrap piece of conduit subtracted
      from (distance "A" to "C" plus distance "C" to "D") is the Shrink,

Note: Both time and energy will be saved if conduit can be cut, reamed,
and threaded before bending. The same method can be used on
hydraulic benders.

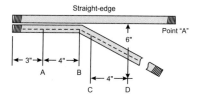

Chicago - Type Bender

Example:

To bend a 6" offset:

1. Make a mark 3' from conduit end. Place conduit in bender with mark at outside edge of jaw.

2. Make three full pumps, making sure handle goes all the way down to the stop.

3. Remove conduit from bender and place alongside straight-edge.

4. Measure 6" from straight-edge to center of conduit. Mark point "D". Use square for accuracy.

5. Mark center of conduit from both directions through bend as shown by broken line. Where lines intersect is point "B".

6. Measure from "A" to "B" to determine distance from "D" to "C", Mark "C" and place conduit In bender with mark at the outside edge of law, and with the kick pointing down, Use a level to prevent dogging conduit,

7. Make three full pumps, making sure handle goes all the way down to the stop,

Note: 1. There are several methods of bending rigid conduit with a Chicago-type bender, and any method that gets the job done in a minimal amount of time with craftsmanship is acceptable.

   2. Whatever method is used, quality will improve with experience.

164

MULTI-SHOT: 90° CONDUIT BENDING

Problem:
A. To measure, thread, cut, and ream conduit before bending.
B. To accurately bend conduit to the desired height of the stub-up (H) and to the desired length of the tail (L).

Given:
A. Size of conduit = 2"
B. Space between conduit (center to center) = 6"
C. Height of stub-up = 36"
D. Length of tail = 48"

Solution:
A. To Determine Radius (R):
   Conduit #1 (inside conduit) will use the minimum radius unless otherwise specified. The minimum radius is eight times the size of the conduit. (See page 167).
   Radius of Conduit #1 = 8 x 2" + 1.25" = 17.25"
   Radius of Conduit #2 = RADIUS #1 + 6" = 23.25"
   Radius of Conduit #3 = RADIUS #2 + 6" = 29.25"

B. To Determine Developed Length (DL):
   Radius x 1.57 = DL
   DL of Conduit #1 = R x 1.57 = 17.25" x 1.57 = 27"
   DL of Conduit #2 = R x 1.57 = 23.25" x 1.57 = 36.5"
   DL of Conduit #3 = R x 1.57 = 29.25" x 1.57 = 46"

C. To Determine Length of Nipple:
   Length of Nipple, Conduit #1 = L + H + DL – 2R
                                = 48" + 36" + 27" – 34.5"
                                = 76.5"
   Length of Nipple, Conduit #2 = L + H + DL – 2R
                                = 54" + 42" + 36.5" – 46.5"
                                = 86"
   Length of Nipple, Conduit #3 = L + H + DL – 2R
                                = 60" + 48" + 46" – 58.5"
                                = 95.5"

Notes: 1. For 90° bends, shrink = 2R – DL
         2. For offset bends, shrink = Hypotenuse – Side Adjacent

Layout and Bending:

A. To locate point "B," measure from point "A," the length of the stub-up minus the radius. On all three conduit, point "B" will be 18.75" from point "A." (See page 167.)

B. To locate point "C," measure from point "D," the length minus the radius, (see page 164). On all three conduit, point "C" will be 30.75" from point "D." (See page 167.)

C. Divide the developed length (point "B" to point "C") into equal spaces. Spaces should not be more than 1.75" to prevent wrinkling of the conduit. On Conduit #1, 17 spaces of 1.5882" each would give us 18 shots of 5° each. Remember there is always one less space than shot. When determining the number of shots, choose a number that will divide into 90 an even number of times.

D. If an elastic numbered tape is not available, try the method illustrated.

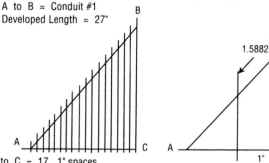

A to B = Conduit #1
Developed Length = 27"

A to C = 17 1" spaces
A to B = 17 1.5882" spaces
C = table or plywood corner

Measure from Point "C" (table corner) 17 inches along table edge to Point "A" and mark. Place end of rule at Point "A." Point "B" will be located where 27" mark meets table edge B–C. Mark on board, then transfer to conduit.

MULTI-SHOT: 90° CONDUIT BENDING

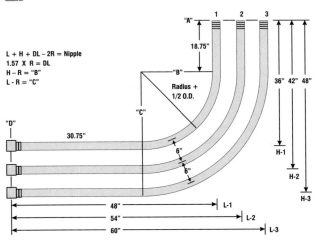

L + H + DL – 2R = Nipple
1.57 X R = DL
H – R = "B"
L – R = "C"

	To Locate Point "B"					To Locate Point "C"			
H#1	-	Radius #1	=	"B"	L#1	-	Radius #1	=	"C"
36-	-	17.25'	=	"B"	48'	-	17.25'	=	"C"
		18.75'	=	"B"			30.75'	=	"C"
H#2	-	Radius #2	=	"B"	L#2	-	Radius #2	=	"C"
42'	-	23.25'	=	"B"	54"	-	23.25'	=	"C"
		18.75'	=	"B"			30.75'	=	"C"
H#3	-	Radius #3	=	"B"	L#3	-	Radius #3	=	"C"
48'	-	29.25'	=	"B"	60'	-	29.25'	=	"C"
		18.75'	=	"B"			30.75'	=	"C"

Points "B" and "C" are the same distance from the end on all three conduits.

⚡ OFFSET BENDS

EMT: Using Hand Bender

An offset bend is used to change the level, or plane, of the conduit. This is usually necessitated by the presence of an obstruction in the original conduit path.

Step One:

Determine the offset depth (X).

Step Two:

Multiply the offset depth "X" the multiplier for the degree of bend used to determine the distance between bends.

Angle		Multiplier
10° x 10°	=	6
22½° x 22½°	=	2.6
30° x 30°	=	2
45° x 45°	=	1.4
60° x 60°	=	1.2

Example: If the offset depth required (X) is 6", and you intend to use 30° bends, the distance between bends is 6" x 2 = 12".

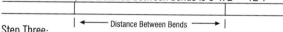

|←———— Distance Between Bends ————→|

Step Three:

Mark at the appropriate points, align the arrow on the bender with the first mark, and bend to desired degree by aligning EMT with chosen degree line on bender.

Step Four:

Slide down the EMT, align the arrow with the second mark, and bend to the same degree line. Be sure to note the orientation of the bender head. Check alignment.

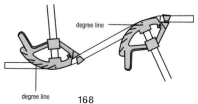

degree line

degree line

🔧 90° BENDS

EMT: Using Hand Bender

The stub-up is the most common bend.

Step One:

Determine the height of the stub-up required and mark on EMT.

Step Two:

Find the "Deduct" or "Take-up" amount from the Take-Up Chart. Subtract the take-up amount from the stub height and mark the EMT that distance from the end.

Step Three:

Align the arrow on bender with the last mark made on the EMT, and bend to the 90° mark on the bender.

Description		Take-Up
½" EMT	=	5"
¾" EMT	=	6"
1"EMT	=	8"
1¼" EMT	=	11"

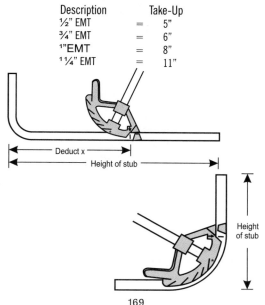

169

 **BACK TO BACK BENDS**

EMT: Using Hand Bender

A back-to-back bend results in a "U" shape in a length of conduit. It's used for a conduit that runs along the floor or ceiling and turns up or down a wall.

Step On:

After the first 90° bend is made, determine the back-to-back length and mark on EMT.

Step Two:

Align this back-to-back mark with the star mark on the bender, and bend to 90°.

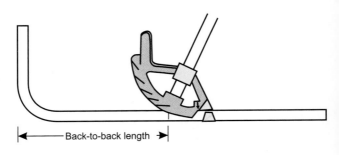

|← ─── Back-to-back length ─→|

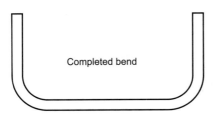

Completed bend

THREE POINT SADDLE BENDS

EMT: Using Hand Bender

The 3-point saddle bend is used when encountering an obstacle (usually another pipe).

Step One:
Measure the height of the obstruction.
Mark the center point on EMT.

Step Two:
Multiply the height of the obstruction by 2.5 and mark this distance on each side of the center mark.

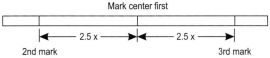

Mark center first

|←— 2.5 x —→|←— 2.5 x —→|

2nd mark   3rd mark

Step Three:
Place the center mark on the saddle mark or notch. Bend to 45°.

Step Four:
Bend the second mark to 22½° angle at arrow.

Step Five:
Bend the third mark to 22½° angle at arrow. Be aware of the orientation of the EMT on all bends. Check alignment.

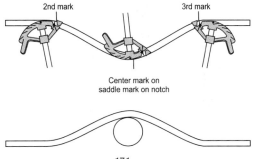

2nd mark   3rd mark

Center mark on
saddle mark on notch

171

🔌 PULLEY CALCULATIONS

The most common configuration consists of a motor with a pulley attached to its shaft, connected by a belt to a second pulley. The motor pulley is referred to as the **Driving Pulley**. The second pulley is called the **Driven Pulley**. The speed at which the Driven Pulley turns is determined by the speed at which the Driving Pulley turns as well as the diameters of both pulleys. The following formulas may be used to determine the relationships between the motor, pulley diameters, and pulley speeds.

> D = **Diameter of Driving Pulley**
> d^1 = **Diameter of Driven Pulley**
> S = **Speed of Driving Pulley** (revolutions per minute)
> s^1 = **Speed of Driven Pulley** (revolutions per minute)

- *To determine the speed of the Driven Pulley (Driven RPM):*

$$s^1 = \frac{D \times S}{d^1} \quad \text{or} \quad \text{Driven RPM} = \frac{\text{Driving Pulley Dia.} \times \text{Driving RPM}}{\text{Driven Pulley Dia.}}$$

- *To determine the speed of the Driving Pulley (Driving RPM):*

$$S = \frac{d^1 \times s^1}{D} \quad \text{or} \quad \text{Driving RPM} = \frac{\text{Driven Pulley Dia.} \times \text{Driven RPM}}{\text{Driving Pulley Dia.}}$$

- *To determine the diameter of the Driven Pulley (Driven Dia.):*

$$d^1 = \frac{D \times S}{s^1} \quad \text{or} \quad \text{Driven Dia.} = \frac{\text{Driving Pulley Dia.} \times \text{Driving RPM}}{\text{Driven RPM}}$$

- *To determine the diameter of the Driving Pulley (Driving Dia.):*

$$D = \frac{d^1 \times s^1}{S} \quad \text{or} \quad \text{Driving Dia.} = \frac{\text{Driven Pulley Dia.} \times \text{Driven RPM}}{\text{Driving RPM}}$$

USEFUL KNOTS

Bowline

Running bowline

Bowline on the bight

Clove hitch

Sheep shank

Rolling hitch

Single
blackwall hitch

Catspaw

Double
blackwall hitch

Square knot

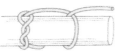

Timber hitch
with half hitch

Single
sheet bend

 HAND SIGNALS

Stop

Dog everything

Emergency stop

Travel

Travel both tracks (crawler cranes only)

Travel one track (crawlers)

Retract boom

Extend boom

Swing boom

174

🔧 HAND SIGNALS

Raise load

Lower load

Main hoist

Move
slowly

Raise boom
and lower load
(flex fingers)

Lower boom
and raise load
(flex fingers)

Use
whip line

Boom up

Boom
down

ELECTRICAL SAFETY DEFINITIONS

Note: Some NFPA 70E definitions include informational notes, which are shown below. Comments shown in italics under some definitions are additional explanations that do not appear in NFPA 70E.

Arc-Flash Hazard: A source of possible injury or damage to health associated with the possible release of energy caused by an electric arc.

Informational Note No. 1: The likelihood of occurrence of an arc flash incident increases when energized electrical conductors or circuit parts are exposed or when they are within equipment in a guarded or enclosed condition, provided a person is interacting with the equipment in such a manner that could cause an electric arc. An arc flash incident is not likely to occur under normal operating conditions when enclosed energized equipment has been properly installed and maintained.

Boundary, Arc Flash: When an arc flash hazard exists, an approach limit from an arc source at which incident energy equals 1.2 cal/cm² (5 J/cm²).

Boundary, Limited Approach: An approach limit at a distance from an exposed energized electrical conductor or circuit part within which a shock hazard exists.

Boundary, Restricted Approach: An approach limit at a distance from an exposed energized electrical conductor or circuit part within which there is an increased likelihood of electric shock, due to electrical arc-over combined with inadvertent movement.

De-energized: Free from any electrical connection to a source of potential difference and from electrical charge; not having a potential different from that of the earth.

Comment: This is a key concept of NFPA 70E. The safest way to work on electrical conductors and equipment is de-energized. See Electrically Safe Work Condition.

Electrical Hazard: A dangerous condition such that contact or equipment failure can result in electrical shock, arc-flash burn, thermal burn, or arc-blast injury.

Electrical Safety: Identifying hazards associated with the use of electrical energy and taking precautions to reduce the risk associated with those hazards.

⚡ ELECTRICAL SAFETY DEFINITIONS

Note: Some NFPA 70E definitions include informational notes, which are shown below. Comments shown in italics under some definitions are additional explanations that do not appear in NFPA 70E.

Electrically Safe Work Condition: A state in which an electrical conductor or circuit part has been disconnected from energized parts, locked/tagged in accordance with established standards, tested to verify the absence of voltage, and, if necessary, temporarily grounded for personnel protection.

Comment: This is a key concept of NFPA 70E. The safest way to work on electrical conductors and equipment is de-energized. The process of turning off the electricity, verifying that it is off, and ensuring that it stays off while work is performed is called "establishing an electrically safe work condition." Many people call the process of ensuring that the current is removed "lockout/tagout"; however, lockout/tagout is only one step in the process.

Energized: Electrically connected to, or is, a source of voltage.

Incident Energy: The amount of thermal energy impressed on a surface, a certain distance from the source, generated during an electrical arc event. Incident energy is typically expressed in calories per square centimeter (cal/cm^2).

Qualified Person: One who has demonstrated skills and knowledge related to the construction and operation of electrical equipment and installations and has received safety training to identify the hazards and reduce the associated risk.

Comment: A person can be considered qualified with respect to certain equipment and methods, but still be unqualified for others. Holding a license or "having done it before" does not make a person qualified. The individual must meet the NFPA 70E definition of qualified person for the specific task being performed.

Working Distance: The distance between a person's face and chest area and a prospective arc source.

Informational Note: Incident energy increases as the distance from the arc source decreases.

WHO IS RESPONSIBLE FOR ELECTRICAL SAFETY?

NFPA 70E, like OSHA, states that both employers and employees are responsible for preventing injury.

- Employers shall provide safety-related work practices and shall train the employees.

- Employees shall implement the safety-related work practices established.

- Multiple employers often work together on the same construction site or in buildings and similar facilities. Some might be onsite personnel working for the host employer, while others are "outside" personnel such as electrical contractors, mechanical and plumbing contractors, painters, or cleaning crews. Outside personnel working for the host employer are employees of contract employers.

- NFPA 70E requires that when a host employer and contract employer work together within the limited approach boundary or the arc-flash boundary of exposed energized electrical conductors or circuit parts, they must coordinate their safety procedures.

- Where the host employer has knowledge of hazards covered by NFPA 70E that are related to the contract employer's work, there shall be a documented meeting between the host employer and the contract employer.

- Outside contractors often are required to follow the host employer's safety procedures.

- Multiple employers involved in the same project sometimes decide to follow the most stringent set of safety procedures.

- Whichever approach is taken, the decision should be recorded in the safety meeting documentation. In accordance with NFPA 70E, where the host employer has knowledge of hazards covered by 70E that are related to the contract employer's work, there shall be a documented meeting between the host employer and the contract employer.

LOCKOUT-TAGOUT AND ELECTRICALLY SAFE WORK CONDITION

The term lockout/tagout refers to specific practices and procedures to safeguard employees from the unexpected energization or startup of machinery and equipment, or the release of hazardous energy during service or maintenance activities. OSHA and NFPA 70E address the control of hazardous energy during service or maintenance of machines or equipment.

OSHA's standard for The Control of Hazardous Energy (Lockout/Tagout), found in Title 29 of the Code of Federal Regulations (CFR) Part 1910.147, addresses the practices and procedures necessary to disable machinery or equipment, thereby preventing the release of hazardous energy while employees perform servicing and maintenance activities. Other OSHA standards, such as 29 CFR 1910.269 and 1910.333 also contain energy control provisions.

Article 120 in NFPA 70E contains requirements for lockout/tagout as well as procedures for establishing and verifying an electrically safe work condition.

Establishing and verifying an electrically safe work condition shall include all of the following steps, which shall be performed in the order presented, if feasible:

1. Identify the power sources.

2. Disconnect power sources.

3. If possible, visually verify that power is disconnected.

4. Release stored electrical energy.

5. Block or relieve stored nonelectrical energy.

6. Apply lockout/tagout devices.

7. Test for the absence of voltage.

8. Install temporary protective grounding equipment if there is a possibility of induced voltages or stored electrical energy.

ELECTRICAL SAFETY: SHOCK PROTECTION BOUNDARIES

130.4(E)(a) Approach Boundaries for Shock Protection, Alternating-Current Systems

Phase-to-Phase Voltage	Limited Approach Boundary Movable	Limited Approach Boundary Fixed	Restricted Approach Boundary
Less than 50	Not specified	Not specified	Not specified
50-150	10 ft	3 ft 6 in.	Avoid contact
151-750	10 ft	3 ft 6 in.	1 ft
751-15000	10 ft	5 ft	2 ft 2 in.
15001-36000	10 ft *	6 ft	2 ft 9 in.
36001-46000	10 ft	8 ft	2 ft 9 in.
46001-72500	10 ft	8 ft	3 ft 6 in.
72600-121000	10 ft 8 in.	8 ft	3 ft 6 in.

* This includes circuits where the exposure does not exceed 120 volts nominal.

Adapted from NFPA 70E®, *Standard for Electrical Safety in the Workplace*®, Table 130.4(E)(a) with permission from National Fire Protection Association, Quincy, MA 02169. This reprinted material is not the complete and official position of the NFPA or the referenced subject, which is represented only by the standard in its entirety.

Where approaching personnel are exposed to energized electrical conductors or circuit parts, the approach boundaries are as follows:

- **Limited Approach Boundary:** This boundary is an approach limit at a distance from an exposed energized electrical conductor or circuit part within which a shock hazard exists. This boundary is larger for movable conductors than for fixed circuit parts.

- **Restricted Approach Boundary:** This boundary is an approach limit at a distance from an exposed energized electrical conductor or circuit part within which there is an increased likelihood of electric shock, due to electrical arc-over combined with inadvertent movement. It allows for the fact that a person's hand or tool might slip, or someone else might jostle the worker from behind.

INFORMATION USUALLY FOUND ON AN ARC-FLASH EQUIPMENT LABEL

(Courtesy of Charles R. Miller)

Electrical equipment such as switchboards, panelboards, industrial control panels, meter socket enclosures, and motor control centers that are in other than dwelling units, and are likely to require examination, adjustment, servicing, or maintenance while energized, shall be field marked with a label containing the information in NFPA70E, 130.5(H).

The available incident energy at the working distance. Instead of the available incident energy and the corresponding working distance, the arc-flash PPE category could have been on this label. See NFPA 70E, 130.5(H).

When an arc flash hazard exists, this is the distance from an arc source at which incident energy equals 1.2 cal/cm2. The onset of a second-degree burn is assumed to be when the skin receives 1.2 cal/cm2 of incident energy

When incident energy is on the label, it is based on a working so the working distance has to be on the label as well.

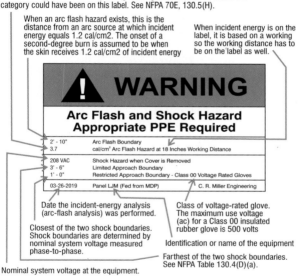

Date the incident-energy analysis (arc-flash analysis) was performed.

Class of voltage-rated glove. The maximum use voltage (ac) for a Class 00 insulated rubber glove is 500 volts

Closest of the two shock boundaries. Shock boundaries are determined by nominal system voltage measured phase-to-phase.

Identification or name of the equipment

Farthest of the two shock boundaries. See NFPA Table 130.4(D)(a).

Nominal system voltage at the equipment.

ELECTRICAL SAFETY: PERSONAL PROTECTION EQUIPMENT GUIDE

Required Personal Protective Equipment (PPE)	
PPE Category	**PPE**
1	**Arc-Rated Clothing, Minimum Arc Rating of 4 cal/cm² (16.75 J/cm²)[a]** Arc-rated long-sleeve shirt and pants or arc-rated coverall Arc-rated face shield[b] or arc flash suit hood Arc-rated jacket, parka, high-visibility apparel, rainwear, or hard hat liner (AN)[f] **Protective Equipment** Hard hat Safety glasses or safety goggles (SR) Hearing protection (ear canal inserts)[c] Heavy-duty leather gloves, or rubber insulating gloves with leather protectors (SR)[d] Leather footwear[e] (AN)
2	**Arc-Rated Clothing, Minimum Arc Rating of 8 cal/cm² (33.5 J/cm²)[a]** Arc-rated long-sleeve shirt and pants or arc-rated coverall Arc-rated flash suit hood or arc-rated face shield[b] and arc-rated balaclava Arc-rated jacket, parka, high-visibility apparel, rainwear, or hard hat liner (AN)[f] **Protective Equipment** Hard hat Safety glasses or safety goggles (SR) Hearing protection (ear canal inserts)[c] Heavy-duty leather gloves, arc-rated gloves, or rubber insulating gloves with leather protectors (SR)[d] Leather footwear[e]
3	**Arc-Rated Clothing Selected So That the System Arc Rating Meets the Required Minimum Arc Rating of 25 cal/cm² (104.7 J/cm²)[a]** Arc-rated long-sleeve shirt (AR) Arc-rated pants (AR) Arc-rated coverall (AR) Arc-rated arc flash suit jacket (AR) Arc-rated arc flash suit pants (AR) Arc-rated arc flash suit hood Arc-rated gloves or rubber insulating gloves with leather protectors (SR)[d] Arc-rated jacket, parka, high-visibility apparel, rainwear, or hard hat liner (AN)[f] **Protective Equipment** Hard hat Safety glasses or safety goggles (SR) Hearing protection (ear canal inserts)[c] Leather footwear[e]

(continued on next page)

⏻ ELECTRICAL SAFETY: PERSONAL PROTECTION EQUIPMENT GUIDE

PPE Category	PPE
4	**Arc-Rated Clothing Slected so That the System Arc Rating Meets the Required Minimum Arc Rating of 40 cal/cm² (167.5 J/cm²)[a]** Arc-rated long-sleeve shirt (AR) Arc-rated pants (AR) Arc-rated coverall (AR) Arc-rated arc flash suit jacket (AR) Arc-rated arc flash suit pants (AR) Arc-rated arc flash suit hood Arc-rated gloves or rubber insulating gloves with leather protectors (SR)[d] Arc-rated jacket, parka, high-visibility apparel, rainwear, or hard hat liner (AN)[f] **Protective Equipment** Hard hat Safety glasses or safety goggles (SR) Hearing protection (ear canal inserts)[c] Leather footwear[e]

AN: as needed (optional). AR: as required. SR: selection required.

[a]*Arc rating is* defined in Article 100.
[b]Face shields are to have wrap-around guarding to protect not only the face but also the forehead, ears, and neck, or, alternatively, an arc-rated arc flash suit hood is required to be worn.
[c]Other types of hearing protection are permitted to be used in lieu of or in addition to ear canal inserts provided they are worn under an arc-rated arc flash suit hood.
[d]Rubber insulating gloves with leather protectors provide arc flash protection in addition to shock protection. Higher class rubber insulating gloves with leather protectors, due to their increased material thickness, provide increased arc flash protection.
[e]Footwear other than leather or dielectric shall be permitted to be used provided it has been tested to demonstrate no ignition, melting, or dripping at the minimum arc rating for the respective arc flash PPE category.
[f]The arc rating of outer layers worn over arc-rated clothing as protection from the elements or for other safety purposes, and that are not used as part of a layered system, shall not be required to be equal to or greater than the estimated incident energy exposure.

ALTERNATIVE ENERGY

Distributed generation systems are designed to work either independently or in parallel with the electric utility grid and have the goal of reducing utility billing, improving electrical reliability, or selling power back to the utility, and being less harmful to the environment. There are five basic types of distributed generation systems: engine-generation systems, solar photovoltaic systems, wind turbines, fuel cells, and microturbines.

Engine-Generation Systems

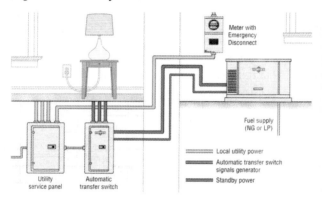

Engine-generation is the most common type of distributed generation system currently available and can be used almost anywhere. Engine-generators have the following:

- An internal combustion engine that runs on a variety of fuels.

- Components that consist of the engine and either an induction generator or a synchronous generator.

- An engine that is either a standby rated or a prime rated. A standby-rated engine is rated to deliver power for the duration of a utility outage. A prime-rated engine is rated to deliver a continuous output with approximately 10% reserved for surges.

Solar Photovoltaic Systems

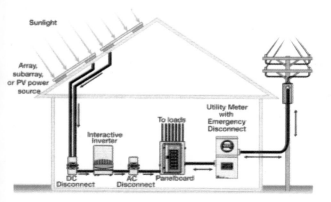

Solar photovoltaic power converts sunlight to dc electrical energy. Solar is limited because of its requirement of sunlight.

- The operation of a solar system is automatic.

- Components consist of foundation and supports, either fixed or tracking arrays, and one or more inverters.

- Per the *NEC* 690.4(C), equipment and all solar-associated wiring and interconnections shall be installed by qualified persons only.

- The PV system disconnecting means shall be installed at a readily accessible location.

ALTERNATIVE ENERGY

Wind Turbines

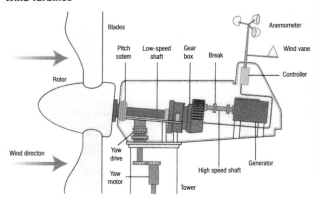

Wind power converts wind to either ac or dc electrical energy. Wind is limited because it needs to be in an area of steady reliable wind.

- Wind is useful as a supplemental power source, but not as a backup source.

- The components of wind power are self-contained wind turbines and support towers.

- The turbine generator can be either directly connected to the fan blade or by a gearbox.

- See *NEC* Article 694 for more information regarding wind (turbine) electric systems.

Fuel Cells

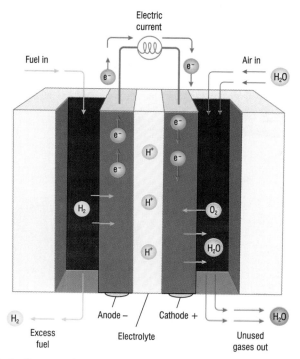

Fuel cells use an electrochemical reactor to generate dc electrical energy. Fuel cells are basically batteries that use hydrogen and oxygen as fuel instead of storing electrical energy.

- A fuel cell generator has no moving parts.

- The fuel cell is composed of a fuel processor, individual fuel cells, fuel cell stack, and power-conditioning equipment.

187

⚡ ALTERNATIVE ENERGY

- Fuel cells can have extremely high operating temperatures, which can limit where they can be used.

- The fuel processor converts hydrocarbon fuel into a relatively pure hydrogen gas.

- Fuel cell systems cannot be installed in a Class I hazardous location (see *NEC* Article 501) and should be installed outside where possible. See *NEC* Article 692 for more information on fuel cell systems. See *NEC* 700.12(E), 701.12(E), and 708.20(H) for additional information regarding fuel cell systems and their requirements.

Microturbines

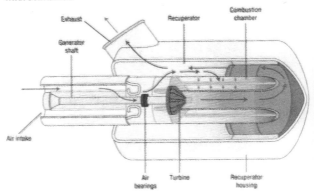

Microturbines are small, single-staged combustion turbines. They can generate either ac or dc electrical energy. Microturbines are also limited in where they can be used due to their high operating temperatures.

- Microturbines range in size from 25 to 500 kW and are modular. They can operate on a wide variety of fuels but are only considered as a renewable energy source.

- The components of microturbines are the compressor, combustion chamber, turbine, generator, recuperator, and the power controller.

ALTERNATIVE ENERGY

- Microturbines are capable of being a stand-alone unit, but the generator loading needs to be relatively steady due to the microturbines' inability to respond quickly.

Interconnected Generation Systems

Interconnected generation systems are one of two basic types: passive or active. Passive generation technologies have no control over power production (wind and solar). Active generation technologies have control over power production and can be regulated to load demands. All grid-connected generation systems must comply with *NEC* Article 705 and with IEEE 1547, *Standard for Interconnection and Interoperability of Distributed Energy Resources with Associated Electric Power Systems Interfaces*.

Utility-interactive power inverters regulate the conversion of dc power into 60 Hz ac voltage waveform in parallel with another ac source (e.g., the electric grid). These systems should comply with *NEC* Articles 690–692, and 705, as required.

Distributed generation systems that are capable of being connected to the grid must have a disconnecting means capable of disconnecting from the grid to prevent the potential hazard of back-feed. See *NEC* 404.6(C).

JUNCTION BOX SIZING

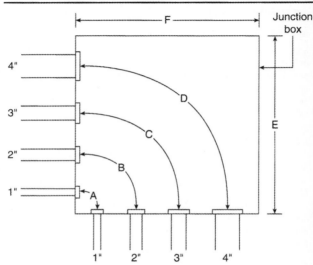

A—6 times conduit size = 6" minimum
B—6 times conduit size = 12" minimum
C—6 times conduit size = 18" minimum
D—6 times conduit size = 24" minimum
E—6 times largest conduit size, plus
    all other conduits entering.
    6 x 4 + 3 + 2 + 1 = 30" minimum
F—In this case, same as E.

SELECTING AND USING TEST INSTRUMENTS

Single Instrument

A multimeter combines the voltmeter, ohmmeter, and milliammeter into a single instrument. In the field, this one instrument can be used to measure (alternating current) ac and (direct current) dc voltages, ac and dc current flow, and electrical resistance.

Although analog multimeters have long been available, most multimeters used today are digital multimeters (DMMs).

Analog:

Digital:

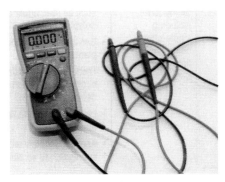

SELECTING AND USING TEST INSTRUMENTS

Instrument Safety Categories

Overvoltage Categories for Test Instruments

Overvoltage Category	In Brief	Examples
CAT IV	Three-phase at utility connection, any outdoor conductors	• Refers to the "origin of installation": i.e., where low-voltage connection is made to utility power • Electricity meters, primary overcurrent protection equipment • Outside and service entrance, service drop from pole to building, run between meter and panel • Overhead line to detached building, underground line to well pump
CAT III	Three-phase distribution, including single-phase commercial lighting	• Equipment in fixed installation, such as switchgear and polyphase motor • Bus and feeder in industrial plants • Feeders and short branch circuits, distribution panel devices • Lighting system in larger buildings • Appliance outlets with short connection to service entrance

SELECTING AND USING TEST INSTRUMENTS

Overvoltage Category	In Brief	Examples
CAT IV	Single-phase receptacle connected loads	• Appliance, portable tool and other household and similar loads
		• Outlet and long branch circuits
		• Outlets at more than 10 meters (30 ft) from CAT III source
		• Outlets at more than 20 meters (60 ft) from CAT IV source
CAT I	Electronic	• Protected electronic equipment
		• Equipment connected to (Source) circuits in which measures are taken to limit transient overvoltages to an appropriately low level
		• Any high-voltage, low-energy source derived from a highwinding resistance transformer, such as the high-voltage section of a copier

Reproduced with permission, Fluke Corporation.

📖 SELECTING AND USING TEST INSTRUMENTS

Selecting an Appropriate Multimeter

- Choose a meter that is properly rated for the circuit or component to be tested.

- Ensure that both the meter and the leads have the correct category rating for each task.

- Check that the ratings of the leads or accessories meet or exceed the rating of the multimeter.

- Ensure that the meter case is not wet, oily, or cracked; that the input jacks are not broken; and that there are no other obvious signs of damage.

- Check the test leads carefully. Ensure that the insulation is not cut, cracked, or melted and that the tips are not loose.

Testing a Multimeter

To measure continuity:

- Set the multimeter to the lowest setting for resistance.

- Touch the tips of the two probes together.

- The display should show 0 ohms (Ω). A DMM will typically display OL (overload or out of limits) when a circuit is open (lacks continuity).

- If OL is displayed during this test, either the meter is defective or the test leads are defective.

To check a meter's ability to measure resistance:

- Use a resistor with a known value.

- Set the multimeter to measure resistance.

- Place one probe at each end of the resistor.

SELECTING AND USING TEST INSTRUMENTS

- The value in the display should be very close to the known value of the resistor being used for the test.

- If the multimeter does not measure the resistance correctly, double-check it by repeating the test with a different resistor of a known value.

To check the dc voltage function:

- Use a new battery.

- Set the multimeter to measure dc voltage.

- Place the red probe on the positive terminal of the battery and the black probe on the negative terminal.

- The multimeter should display a reading of, or close to, the battery's labeled voltage.

- Reversing the leads on a digital meter will display a negative (−) reading.

To check the ac-voltage function:

- Use a 120 VAC receptacle known to be energized.

- Plug in and turn on a lamp or another simple device to ensure that the receptacle is energized. Set the multimeter to measure ac voltage.

- Place the red probe into the energized side—the smaller slot of the receptacle—and the black probe into the other side (neutral).

- The multimeter should display a reading of, or close to, 120 VAC.

If the multimeter fails any one of these tests, it is best to return it to the manufacturer for service. Technicians should never attempt to open and repair the meter's electronics or movement. If the multimeter will not perform any functions or provide a display when tested, the problem may be weak or dead batteries. DMMs have a battery indicator that will display when the battery is weak but still functional. Always replace the battery as soon as the indication is displayed to ensure continued reliable operation.

If the batteries are not the problem, determine if there is an accessible fuse inside the battery compartment. If there is, check to see whether it is open; another meter may be needed to check the continuity of the fuse if its condition cannot be determined visually. Replace an open fuse with one of the proper current and voltage rating. Using the wrong fuse, especially one that has a higher current rating than it should, can result in the meter being damaged beyond repair.

Return a multimeter to the manufacturer for an annual calibration to ensure its accuracy.

Common Testing Errors

Try to avoid the following common errors when using a multimeter to measure voltage:

- Measuring voltage while the red test lead is in the wrong jack. The test-lead jack for the red lead is usually marked with the V symbol and is often marked with the Ω symbol. Applying voltage to the meter while the red lead is plugged into the wrong jack may damage the meter.

- Measuring ac voltage on a dc setting. The dial symbol usually associated with measuring ac voltage is a V with a wavy line above it, or an mV with the same wavy line above. The latter is for measuring in the millivolt range.

- Keeping the test probe in contact with an energized surface longer than necessary. The longer it is in contact, the more time there is for an accident to occur.

- Using the meter above its rated voltage. Technicians should have a general idea of the expected voltage or current to be measured and ensure that the meter is capable of safely making such readings.

SELECTING AND USING TEST INSTRUMENTS

Using a Multimeter to Measure Voltage

Take all available and necessary precautions to avoid contact with energized surfaces, and ensure that the probe of the test lead does not accidentally bridge two points with different electrical potentials. Use test leads with finger guards, and always keep fingers behind them.

- Select an appropriate multimeter for the job.

- Visually inspect it and the test leads for any signs of damage. Before using the multimeter on the job, always test it on a known voltage source to verify that it is functioning properly.

- Insert the black test lead into the common input jack (COM) and the red test lead into the input jack for ac volts (V or VΩ). Take care to insert the test leads into the correct jacks.

- Select ac as the type of voltage to be measured. If the meter has no auto-ranging feature, select the voltage range.

- Place the red probe onto the energized side of the circuit.

- Place the black probe onto the neutral side of the circuit or to ground.

- Read the voltage displayed on the meter while both probes make good electrical contact with their individual targets.

Using a Multimeter to Measure Resistance

- The ohmmeter function can be used to determine whether a circuit has continuity or to determine a specific resistance value.

- Select an appropriate multimeter for the job. Visually inspect it and the test leads for any signs of damage.

- Test the multimeter to verify that it is functioning properly and can measure continuity and resistance. A small resistor of a known value can be used to verify meter operation.

SELECTING AND USING TEST INSTRUMENTS

- Insert the black test lead into the common input jack (COM) and the red test lead into the input jack for VΩ or Ω. If the meter has no auto-ranging feature, select the voltage range, starting with the highest.

- Set the meter to measure resistance (Ω).

- Ensure the circuit is deenergized.

- Touch the test leads to the two points in the circuit across the resistance.

- Read and record the resistance value displayed on the meter.

Using a Multimeter to Measure Current

A technician will frequently use a meter to measure ac. In some cases, ac values that are in the milliamp, or even the microamp (μA), range may need to be measured.

Most multimeters can measure these small current values by placing the meter in series with the circuit and then energizing the circuit to read the current flow.

Multimeters typically have a maximum current limit of 10 A when using this method of measurement. Do not exceed the maximum current limit of the meter.

Remember that exceeding the current limit with the meter in series with the load can result in significant damage to the meter.

To measure current values that exceed the meter's limit, use either a clamp-on ammeter or a clamp-on ammeter accessory designed for the multimeter in use.

Selecting a Clamp-On Ammeter

Many clamp-on ammeters are designed to provide most or all the functions of a DMM, with the added convenience of a built-in ammeter that can measure a significant amount of current. Therefore, they may be used to measure voltage, resistance, and current like any DMM. A major distinguishing feature of a clamp-on ammeter is that it has a current transformer built into the jaws that can be opened and closed around a conductor. Closing the jaws around a conductor enables the current flow to be measured without having to handle the conductor or disrupt the operation of the component or system being tested. Through induction, a small current is induced in the jaws by the current in the conductor.

Typical clamp-on ammeter features and functions include, but are not limited to, the following:

- A selector switch for selecting the desired test function (voltage, current, or resistance)

- An auto-ranging feature to automatically select the proper measurement range

- A HOLD function to freeze the reading shown on the display

- A minimum/maximum memory function to determine the highest and lowest reading over the course of a test. (The maximum function is valuable for measuring the inrush current when a motor starts.)

- A capacitor-testing feature that measures capacitance and checks for shorts and opens

- A continuity beeper that is activated when continuity through a circuit is detected. (The beeper is an advantage for troubleshooting, particularly in tight spaces, because attention can remain focused on the meter leads—listening for the beeper instead of diverting attention from the leads to check the readings.)

- Overload protection to prevent damage to the meter and to protect the user

Always read and follow the manufacturer's instructions for the meter to make the best use of its features.

SELECTING AND USING TEST INSTRUMENTS

Carefully review an ammeter's specifications, features, and functions. Be sure that it has overload protection to protect the meter and the user. Choose an ammeter that can provide the degree of accuracy required. Ensure the ammeter is properly rated for the expected current. Inductive loads, such as electric motors, experience a significant inrush of current when they start. This surge occurs when a motor is first energized and must be brought up to speed from a complete stop. Although the surge may last less than one second, the current can be many times higher than the current shown on the motor data plate, especially if the motor has mechanical or electrical defects. Remember to consider the possible inrush current when selecting an ammeter and the range. The range can be changed, if needed, to a more appropriate level while the ammeter is actively measuring current. Meters with an auto-ranging feature eliminate this concern. Ensure that the ammeter has the correct category rating for each task. Also, be sure that the ratings of the leads or other accessories meet or exceed the rating of the ammeter. When making a current measurement, ensure that the clamp is firmly and completely closed to get an accurate reading.

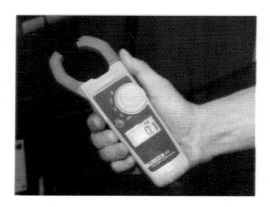

SELECTING AND USING TEST INSTRUMENTS

Testing a Clamp-On Ammeter

- Verify that the clamp-on ammeter is working properly before using it. The procedures for testing a clamp-on ammeter are the same as those used to test the functions of a multimeter.

- Always inspect a clamp-on ammeter before using it. Ensure that the case is not cracked or greasy and that there are no obvious signs of damage. Be sure that the jaw tips are not dirty and that they meet and interlock properly. Proper jaw alignment at the tips is essential for an accurate measurement. If the jaws are dirty or misaligned, the meter will not read correctly.

- Test a clamp-on ammeter to verify that it can measure continuity/resistance and voltage. Do not use the meter if it fails any one of the tests. If the ammeter is not performing any functions at all when it is tested, check for a blown fuse and/or replace the battery.

Using a Clamp-On Ammeter to Measure Current

Most digital clamp-on ammeters have an auto-ranging feature. If, however, the range needs to be set manually, always start at the highest range and adjust the range as needed to obtain an accurate reading. Place the jaws of the ammeter around only one conductor at a time. Placing the jaws around two conductors at the same time will produce an inaccurate reading.

- While the system is deenergized, select the location for the current measurement and separate the target conductor from others so that the jaws can safely and easily be snapped closed around the conductor when the power is on.

- Make any preparations for measurement that can be made while the power is off to significantly reduce exposure to serious injury.

- Note that wire insulation is not a factor; there is no need to place the jaws around bare conductors. In fact, doing so can be dangerous.

- If the current reading is expected to be at the low end of the lowest ammeter range, a more accurate reading can be obtained by wrapping the single conductor wire multiple times around one jaw of the ammeter.

- It is not often necessary for digital meters, but it is a significant help to read low current values on an analog model. Each time the wire passes through the jaws, the current it carries is sensed by the meter.

- If the same conductor passes through the jaws five times, the current measured will be five times the actual current flow. Be sure, however, to divide the reading by the number of times the wire passes through the jaws.

- While wrapping the conductor, do not allow the wraps to cross over each other. This can affect the accuracy.

- Note that this works equally well with a clamp-on ammeter accessory used with multimeters.

Non-contact Voltage Tester

🔌 SELECTING AND USING TEST INSTRUMENTS

Although a non-contact voltage tester is not suitable for work that requires any level of accuracy, it is useful for quickly verifying that a circuit is energized or deenergized. It is important to note that it should not replace voltmeter testing to prove that a circuit is deenergized for safety reasons. Electrical safety demands that a more dependable test device be used. The test instrument must be able to test each phase conductor or circuit part both phase-to-phase and phase-to-ground.

Non-contact voltage testers are relatively small, compact, battery-powered instruments that fit in a shirt pocket. They typically have the following features:

- An On/Off power button

- A flashlight or indicator light that shows the tester is operational and the flashlight may help illuminate the workplace (in some models, the flashlight can be operated independently of the voltage tester)

- A high-intensity LED light and/or a warning tone to notify when voltage is detected

- An automatic shutoff to conserve power and extend battery life

- A low-battery indicator

- A built-in self-test feature

Non-contact voltage testers are available in several voltage ranges. A standard model may be sensitive only to voltages above a range of 90–100 VAC.

Dual-voltage models can register and differentiate between standard voltages and lower voltages. This feature makes these models more suitable for testing low-voltage ac circuits for the presence of power.

Note that these tools are not designed to sense the presence of dc voltage, since they rely on the presence of a magnetic field generated by ac. They may occasionally respond to the presence of dc voltage, but not reliably.

🔌 SELECTING AND USING TEST INSTRUMENTS

Selecting a Non-contact Voltage Tester

- When selecting a non-contact voltage tester, choose one with a range suitable for the expected voltage.

- Be sure the instrument is sensitive enough in terms of the minimum voltage that it can detect. Inspect the tester for any obvious signs of damage. Ensure the batteries for the tester are working.

- Many testers offer a battery test button to ensure the device is working properly.

Testing a Non-contact Voltage Tester

Check the tester on a known, energized voltage source prior to using the instrument in the field. Use the built-in self-test feature, if the tester is so equipped. Otherwise, a simple way to test it is to place the tip of the tester near the line-voltage side (smaller slot) of an electrical outlet known to be energized. If the tester is working properly, the tester should give a clear audible and/or visual indication that voltage is present.

Using a Non-contact Voltage Tester

Although non-contact voltage testers may detect voltages up to 1000 VAC or more, working near voltages this high can be extremely hazardous. They are best used to detect voltages well below that value.

When safety is at stake and the intention is to make contact with conductors and/or electrical parts, use a multimeter to conduct the test to ensure that the circuit has been deenergized.

- Select an appropriate non-contact voltage tester for the job.

- Inspect it for any obvious damage.

- Ensure the batteries are working. If the instrument is equipped with a self-test feature, use it to ensure the instrument is functioning properly.

SELECTING AND USING TEST INSTRUMENTS

- Before using the voltage tester on the job, always test it to verify that it is functioning properly.

- Gradually move the tester as close as possible to the wire you wish to test.

- If voltage is present, the tester will react with both an audio and visual signal.

- If the circuit is off, there should be no voltage present and no response from the tester.

- If the circuit is live, the tester should provide a visual, and possibly an audible, indication that voltage is present.

- Move the tester from one power wire to the other to be sure.

- Cycle the power off and watch for the indication that voltage is no longer present.

Notes

Notes

Notes

Notes

Notes

Notes

Notes

Notes

Notes

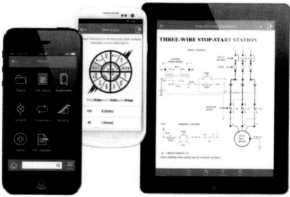